USE OF PERIODIC
SAFETY REVIEW FOR
LONG TERM OPERATION
OF NUCLEAR
POWER PLANTS

The following States are Members of the International Atomic Energy Agency:

AFGHANISTAN	GAMBIA	OMAN
ALBANIA	GEORGIA	PAKISTAN
ALGERIA	GERMANY	PALAU
ANGOLA	GHANA	PANAMA
ANTIGUA AND BARBUDA	GREECE	PAPUA NEW GUINEA
ARGENTINA	GRENADA	PARAGUAY
ARMENIA	GUATEMALA	PERU
AUSTRALIA	GUYANA	PHILIPPINES
AUSTRIA	HAITI	POLAND
AZERBAIJAN	HOLY SEE	PORTUGAL
BAHAMAS	HONDURAS	QATAR
BAHRAIN	HUNGARY	REPUBLIC OF MOLDOVA
BANGLADESH	ICELAND	ROMANIA
BARBADOS	INDIA	RUSSIAN FEDERATION
BELARUS	INDONESIA	RWANDA
BELGIUM	IRAN, ISLAMIC REPUBLIC OF	SAINT KITTS AND NEVIS
BELIZE	IRAQ	SAINT LUCIA
BENIN	IRELAND	SAINT VINCENT AND
BOLIVIA, PLURINATIONAL	ISRAEL	THE GRENADINES
STATE OF	ITALY	SAMOA
BOSNIA AND HERZEGOVINA	JAMAICA	SAN MARINO
BOTSWANA	JAPAN	SAUDI ARABIA
BRAZIL	JORDAN	SENEGAL
BRUNEI DARUSSALAM	KAZAKHSTAN	SERBIA
BULGARIA	KENYA	SEYCHELLES
BURKINA FASO	KOREA, REPUBLIC OF	SIERRA LEONE
BURUNDI	KUWAIT	SINGAPORE
CABO VERDE	KYRGYZSTAN	SLOVAKIA
CAMBODIA	LAO PEOPLE'S DEMOCRATIC	SLOVENIA
CAMEROON	REPUBLIC	SOUTH AFRICA
CANADA	LATVIA	SPAIN
CENTRAL AFRICAN	LEBANON	SRI LANKA
REPUBLIC	LESOTHO	SUDAN
CHAD	LIBERIA	SWEDEN
CHILE	LIBYA	SWITZERLAND
CHINA	LIECHTENSTEIN	SYRIAN ARAB REPUBLIC
COLOMBIA	LITHUANIA	TAJIKISTAN
COMOROS	LUXEMBOURG	THAILAND
CONGO	MADAGASCAR	TOGO
COSTA RICA	MALAWI	TONGA
CÔTE D'IVOIRE	MALAYSIA	TRINIDAD AND TOBAGO
CROATIA	MALI	TUNISIA
CUBA	MALTA	TÜRKİYE
CYPRUS	MARSHALL ISLANDS	TURKMENISTAN
CZECH REPUBLIC	MAURITANIA	UGANDA
DEMOCRATIC REPUBLIC	MAURITIUS	UKRAINE
OF THE CONGO	MEXICO	UNITED ARAB EMIRATES
DENMARK	MONACO	UNITED KINGDOM OF
DJIBOUTI	MONGOLIA	GREAT BRITAIN AND
DOMINICA	MONTENEGRO	NORTHERN IRELAND
DOMINICAN REPUBLIC	MOROCCO	UNITED REPUBLIC
ECUADOR	MOZAMBIQUE	OF TANZANIA
EGYPT	MYANMAR	UNITED STATES OF AMERICA
EL SALVADOR	NAMIBIA	URUGUAY
ERITREA	NEPAL	UZBEKISTAN
ESTONIA	NETHERLANDS	VANUATU
ESWATINI	NEW ZEALAND	VENEZUELA, BOLIVARIAN
ETHIOPIA	NICARAGUA	REPUBLIC OF
FIJI	NIGER	VIET NAM
FINLAND	NIGERIA	YEMEN
FRANCE	NORTH MACEDONIA	ZAMBIA
GABON	NORWAY	ZIMBABWE

The Agency's Statute was approved on 23 October 1956 by the Conference on the Statute of the IAEA held at United Nations Headquarters, New York; it entered into force on 29 July 1957. The Headquarters of the Agency are situated in Vienna. Its principal objective is "to accelerate and enlarge the contribution of atomic energy to peace, health and prosperity throughout the world".

Safety Reports Series No. 121

USE OF PERIODIC SAFETY REVIEW FOR LONG TERM OPERATION OF NUCLEAR POWER PLANTS

INTERNATIONAL ATOMIC ENERGY AGENCY
VIENNA, 2023

© IAEA, 2023

Printed by the IAEA in Austria
October 2023
STI/PUB/2010

IAEA Library Cataloguing in Publication Data

Names: International Atomic Energy Agency.
Title: Use of periodic safety review for long term operation of nuclear power plants / International Atomic Energy Agency.
Description: Vienna : International Atomic Energy Agency, 2023. | Series: IAEA safety reports series, ISSN 1020–6450 ; no. 121 | Includes bibliographical references.
Identifiers: IAEAL 23-01608 | ISBN 978–92–0–117823–7 (paperback : alk. paper) | ISBN 978–92–0–117923–4 (pdf) | ISBN 978–92–0–118023–0 (epub)
Subjects: LCSH: Nuclear power plants — Safety measures. | Nuclear power plants — Maintainability. | Nuclear power plants — Management.
Classification: UDC 621.039.58 | STI/PUB/2010

FOREWORD

A comprehensive assessment of the safety of a nuclear power plant is an important precondition for continuing its operation beyond the originally anticipated time frame. Most Member States consider periodic safety review to be an effective way to obtain an overall view of actual plant safety and determine reasonable and practicable modifications to be made in order to ensure that a high level of safety is maintained during continued operation.

The complex process of conducting a periodic safety review can be aided by appropriately subdividing the tasks into several safety factors. These safety factors are selected based on international experience and are intended to cover all aspects important to the safety of an operating nuclear power plant. In cases where the number of safety factors used and/or their grouping is different because of the specific needs of the operating organization or regulatory body, or owing to particular aspects of the nuclear power plant under review, the comprehensiveness of the periodic safety review can be ensured by other means.

A periodic safety review typically addresses the period until the next periodic safety review or, where appropriate, until the end of planned operation, and considers whether there are any foreseeable circumstances that could endanger the operation of the nuclear power plant in the interim. Continuation of operation of a nuclear power plant beyond the originally anticipated time frame (typically 30–40 years) has become a priority for many operating organizations. Long term operation needs to be justified by safety assessment, with consideration given to the life limiting processes and features of structures, systems and components. As the typical interval between periodic safety reviews is ten years, this means that the third or fourth periodic safety review will possibly evaluate the safety factors related to long term operation, namely design; actual status of structures, systems and components; equipment qualification and ageing.

This publication is intended to provide a common basis and criteria for using periodic safety reviews in support of justification of long term operation or licence renewal, as well as to discuss current challenges, synergies, good practices and examples of corrective actions and safety improvements related to the application of periodic safety reviews for justifying long term operation of nuclear power plants.

The IAEA officers responsible for this publication were A. Duchac, R. Krivanek and G. Petofi of the Division of Nuclear Installation Safety.

CONTENTS

1. INTRODUCTION

1.1. BACKGROUND

Operational nuclear power plants (NPPs) are generally subject to routine reviews of plant operational safety and ad hoc safety reviews in response to operational events. In many Member States, these safety reviews have specific areas of focus and do not typically consider all factors relevant to long term operation (LTO), such as changes in codes, standards and regulations; the cumulative effects of plant ageing; plant modifications; feedback of operating experience; and developments in science and technology.

To capture the aggregate effect of these factors on plant safety, many Member States periodically conduct a comprehensive, integrated safety review, commonly referred to as a 'periodic safety review' (PSR). This type of review relies on a systematic and comprehensive process whereby up to date codes, standards and regulations are considered to provide assurance of the continued viability of the plant's licensing basis, given the cumulative impacts of emerging national and international standards, evolving regulatory requirements, plant ageing, operating experience and technological developments. PSR provides an effective way to obtain an overall view of actual plant safety and the quality of the safety documentation and to determine reasonable and practical modifications to ensure or improve safety.

The main goal of the PSR is to drive the continuous safety improvement of the plant. As the usual interval between PSRs is ten years, this means that the third or fourth PSR is typically used to evaluate the safety factors related to operation beyond the originally planned or licensed lifetime, usually referred to as 'long term operation'. Relevant safety factors include (among others) plant design safety; actual condition of structures, systems and components (SSCs) important to safety; equipment qualification and ageing. The term 'safety factor', introduced in IAEA Safety Standards Series No. SSG-25, Periodic Safety Review for Nuclear Power Plants [1], refers to those aspects of the safety of an operating NPP that are considered important and are not expected to change over the course of LTO. As the LTO assessment and the PSR programme often do not run in parallel, several inputs from LTO to PSR, such as corrective actions and the LTO action plan, are to be fully incorporated into the subsequent PSR.

In February 2020, the IAEA organized a Technical Meeting on Use of Periodic Safety Review in Support of Long Term Operation Safety Assessment. The meeting provided a forum for discussion of current challenges, synergies, good practices and examples of corrective actions and safety improvements related to the application of PSR for justifying LTO for NPPs. The final round table discussion underlined the IAEA's recommendations on the implementation of PSR (reflected in SSG-25 [1]) and LTO (reflected in IAEA Safety Standards Series No. SSG-48, Ageing Management and Development of a Programme for Long Term Operation of Nuclear Power Plants [2]), which are broadly used by Member States. Because the scope and use of PSR in support of the justification of LTO varies among Member States, the operating organizations and regulatory bodies sought IAEA assistance in clarifying the use of PSR to support decision making for the long term. Furthermore, guidance was sought to improve Member States' understanding of the IAEA Safety Standards on PSR and LTO, as well as their adequate implementation.

1.2. OBJECTIVE

The objectives of this publication are:

(a) To provide a common basis and criteria for using PSR results in support of justification of LTO;
(b) To provide information regarding use of results from LTO activities in PSR;
(c) To describe current challenges, synergies and examples of Member States' practices related to the application of PSR to support the LTO of NPPs;
(d) To define good practices for LTO and PSR programmes at NPPs;
(e) To improve understanding of the IAEA Safety Standards on PSR, ageing management and LTO.

This publication is intended for operating organizations, regulatory bodies and their technical support organizations, consultants and advisory bodies.

Guidance and recommendations provided here in relation to identified good practices represent expert opinion but are not made on the basis of a consensus of all Member States.

1.3. SCOPE

This publication addresses the scope of the LTO assessment, methods and applicable criteria to support decision making within the PSR and LTO programmes framework. It is intended to verify whether:

(a) The plant adequately conforms to modern standards and practices;
(b) The (updated) licensing basis will be valid throughout the plant's intended period of operation;

(c)   The arrangements that are in place ensure the maintenance of safe plant operation during LTO;

(d)   The improvements to be implemented address any safety issues that have been identified.

This publication describes in detail:

(a)   Augmentation and amendment of a PSR strategy and methodology when the PSR will be used in support of LTO;

(b)   Coordination and synergies of PSR and LTO activities;

(c)   The function of the PSR global assessment in the justification of LTO;

(d)   Roles and responsibilities in preparation for safe LTO.

The assessment methods described in this publication are not meant to replace other evaluations required by national regulations and national or international standards. Rather, they are intended to provide additional tools for improved coordination and guidance regarding finding potential synergies among the programmes for LTO and PSR.

Due to the nature of the subjects being considered, this publication applies to NPPs. This publication may have a wider applicability, for example, to radioactive waste management facilities, spent fuel storage or handling facilities located on the reactor site. The approaches described in this publication can be adapted, with judgement, to other nuclear installations (e.g. small modular reactors, research reactors).

This publication references several IAEA Safety Standards and Safety Report Series that are needed for cross-referencing purposes while reading this publication. It is strongly recommended that the reader be familiar with Refs [1] and [2] before reading this publication.

1.4. STRUCTURE

This publication consists of seven sections. In addition, there are two annexes illustrating practices in Member States. Following the introduction, Section 2 discusses the PSR strategy and methodology in support of justification of LTO. Section 3 is from the perspective of both an LTO programme and a PSR programme. Section 4 describes the method of global assessment. Section 5 describes the implementation of programmes, commitments and improvements for LTO. Section 6 describes the roles and responsibilities involved in preparation for safe LTO. Section 7 provides a description of the documentation needed in support of justification of

LTO. Annex I provides Member States' experiences and practices. Finally, Annex II provides an example of the steps to be taken in a global assessment.

2. STRATEGY AND METHODOLOGY FOR PERIODIC SAFETY REVIEW IN SUPPORT OF LONG TERM OPERATION JUSTIFICATION

2.1. SCOPE AND CONTENT OF PERIODIC SAFETY REVIEW FOR JUSTIFICATION OF LONG TERM OPERATION

The objective of PSR, as specified in SSG-25 [1], is to determine by means of a comprehensive assessment:

(a) The adequacy and effectiveness of the arrangements and the SSCs (equipment) that are in place to ensure plant safety for:
- The next planned PSR;
- The end of planned operation (if the NPP will cease operation before the next regular PSR);
- The end of LTO (if the PSR is planned to be used in support of justification of LTO).
(b) The extent to which the plant conforms to current national and/or international safety standards and operating practices.
(c) Necessary safety improvements and timescales for the implementation.
(d) The extent to which the safety documentation, including the licensing basis, remains valid.

In general, the PSR outputs could be used in support of justification of LTO in the following ways:

(a) To obtain an overall assessment of actual plant safety;
(b) To confirm that sufficient safety margins of SSCs exist for the next operational period;
(c) To verify the effectiveness of the operating and maintenance programmes for LTO;
(d) To identify any safety concern(s) that require corrective actions;
(e) To determine reasonable safety improvements to enhance plant safety and reliability;
(f) To establish a corrective action plan.

PSR also evaluates the plant organization, processes, programmes and management systems in order to confirm that they are adequate to support the safe operation of the facility.

When PSR is used as an input to decision making for LTO, the documentation needs to clearly define the scope of the assessments related to LTO that are carried out in the PSR. The differences between a standard PSR and a PSR for LTO are as follows:

(a) The PSR scope is extended in accordance with Requirement 16 of IAEA Safety Standards Series No. SSR-2/2 (Rev. 1), Safety of Nuclear Power Plants: Commissioning and Operation [3]. This extended PSR requires:
   — A review of the preconditions for LTO, which covers programmes and documents relevant for ageing management;
   — A review of ageing management and its alignment with SSG-48 [2], especially with regard to scope setting, including special attention to SSCs that are to be included in the scope, in accordance with para. 5.16(b)[1] of SSG-48 [2].
(b) LTO is an extended operation necessitating a feasibility study and a subsequent analysis in terms of financial return, as justification. The results of the assessment of a PSR for LTO of an NPP need to confirm the assumptions of the feasibility study for LTO.
(c) A PSR in support of LTO is a part of the programme for LTO. The volume of physical modifications is typically higher because of objectives related to design improvements, preserving and extending equipment qualification, and addressing technological obsolescence issues.
(d) Intending to use PSR in support of LTO may have a crucial impact on its scheduling, especially if an environmental impact assessment is also required for LTO and if the overall closure of the PSR is needed just before or after entering the period of LTO.
(e) LTO of more than ten years needs to be justified in such a way that the evaluations in the PSR are extended from ten years to the period of LTO.
(f) A review of long term resource planning, preserving competences and knowledge for the entire period of LTO, is included and takes into account

---

[1] "Other SSCs whose failure may prevent SSCs important to safety from fulfilling their intended functions. Examples of such potential failures are:
   — Missile impact from rotating machines;
   — Failures of lifting equipment;
   — Flooding;
   — High energy line break;
   — Leakage of liquids (e.g. from piping or other pressure boundary components)."

the increasing retirement of experienced personnel and possible challenges in recruiting suitably qualified and experienced replacements.

Paragraph 2.10 of SSG-25 [1] states that a PSR can be used "[a]s a systematic safety assessment carried out at regular intervals" and "[i]n support of the decision making process for long term operation". SSG-25 [1] is not specifically intended to provide recommendations for the activities performed during LTO of an NPP. However, the scope of the review of the safety factors can be adapted to facilitate an assessment of the feasibility of LTO.

According to Section 3 of SSG-25 [1], PSR is considered an effective way to obtain an overall view of actual plant safety and to determine reasonable and practicable modifications to be made in order to ensure that a high level of safety is maintained during LTO. The safety improvements identified in the PSR are used as inputs into the decision as to whether to approve LTO (see para. 3.10 of SSG-25 [1]). PSR and LTO programmes need to identify life limiting features of the NPP in order to determine if there is a need to modify, refurbish or replace certain SSCs for the purpose of extending its operating lifetime. Therefore, the scope of the safety factors needs to be adapted to determine the feasibility of LTO. For example, the scope of the safety factor relating to ageing may be expanded to include an evaluation of the safety analyses with time limited assumptions and assessments of ageing effects. In the review, increased importance is given to ageing effects, degradation mechanisms and ageing management programmes (AMPs) [2].

If a combined PSR can be performed for several units at a site, then the results of the PSR could also be used to support an LTO programme covering several units.

2.1.1. Time frame of PSR for LTO

Normally, a PSR covers ten years of operation of an NPP. However, if a PSR is used to justify LTO, the time span considered in the review of the safety factors covers the entire intended period of LTO (see para. 3.7 of SSG-25 [1]). This is typically 20 years, but the time span could be different, depending on national regulations and the plant's intentions. As specified in para. 7.38 of SSG-48 [2], the timespan for reviewing safety factors 2 to 5, which relate to the actual condition of SSCs, equipment qualification, ageing and deterministic safety analysis (DSA), respectively, could be adapted to cover the period of LTO if the PSR is used to support LTO. The operating organization continues to carry out PSR every ten years during the period of LTO (see para. 3.7 of SSG-25 [1]). If an LTO orientated PSR is completed several years (e.g. 2–3 years) prior to the start of the intended period of LTO, some results might no longer be valid.

Additional analyses may be required to confirm the validity of the evaluation results for the justification for LTO. Part of this could be due to obsolescence: a possible solution is to carry out gap analyses within the organization to identify the gaps to be filled.

2.1.2. Scope setting of SSCs

SSG-25 [1] sets the scope of SSCs for conducting PSR at a high level as "SSCs important to safety" and "any non-safety-classified SSCs" whose failure might inhibit or adversely affect a safety function. Paragraph 5.16 of SSG-48 [2] is more specific regarding the selection of in-scope SSCs and identifies three categories of SSCs for inclusion in the scope of ageing management, as follows:

(a)   SSCs important to safety that are necessary to fulfil the fundamental safety functions;
(b)   Other SSCs whose failure may prevent SSCs important to safety from fulfilling their intended safety functions;
(c)   Other SSCs that are credited in the safety analyses (deterministic and probabilistic) as performing the function of coping with certain types of events, consistent with national regulatory requirements.

Table 1 outlines the correlation between the SSCs identified for evaluating safety factors 1–4 in PSR at the high level (i.e. SSCs important to safety and non-safety-classified SSCs whose failure might inhibit or adversely affect safety functions) and the SSCs selected using criteria (a), (b) and (c) above (defined in more detail in SSG-48 [2], para. 5.16) for ageing management review (AMR).

The process of selecting SSCs using the three criteria defined in SSG-48 [2], para. 5.16, for AMR is based on SSCs' safety functions or their intended functions, whereas the process of SSC selection for PSR described in SSG-25 [1] is based on a high level safety classification (i.e. items important to safety and certain items not important to safety). The three criteria defined in SSG-48 [2] provide a more comprehensive and clearer basis for the selection of SSCs than that given in SSG-25 [1].

A comprehensive, well documented and up to date safety case, including safety analysis, is required to adequately complete the scope setting process for LTO. The safety case can also be used to support the PSR of safety factors 1, 5, 6 and 7 relating to plant design and safety analysis.

A clear distinction between those SSCs considered to be within the scope and those out of the scope needs to be defined and documented by the licensee at a structure or component level, including a clear statement about the boundary between the two items, where appropriate.

TABLE 1. CORRELATION OF THE CRITERIA FOR SCOPE SETTING OF SSCs PROVIDED IN SSG-25 [1] AND SSG-48 [2]

In-scope SSCs for safety factors 1–4 of PSR (SSG-25 [1])	Definition	In-scope SSCs for LTO (applicable selection criteria (a), (b) and (c) as defined in para. 5.16 of SSG-48 [2])
Safety systems	A system important to safety, provided to ensure the safe shutdown of the reactor or the residual heat removal from the reactor core, or to limit the consequences of anticipated operational occurrences and design basis accidents [4]	Selection criterion (a) applies directly and includes SSCs delivering the three fundamental safety functions (control of reactivity, cooling and confinement of fissile material)
Safety related items	An item important to safety that is not part of a safety system [4]	Selection criterion (b) applies and includes SSCs whose failure may prevent SSCs important to safety from fulfilling their intended functions. Criterion (b) may include items not important to safety, according to para. 5.51 of SSG-25 [1]
Items not important to safety	Any non-safety classified SSCs whose failure might inhibit or adversely affect safety functions (para. 5.51 of SSG-25 [1])	Several items not important to safety may be included in the scope for LTO based on the selection criteria (b) and (c)
Safety features for design extension conditions	Item that is designed to perform a safety function for, or that has a safety function for, design extension conditions [4]	Selection criterion (c) applies and includes SSCs needed to cope with design extension conditions or to mitigate the consequences of severe accidents

Typically, the list of SSCs important to safety only considers permanent or anchored equipment. Following the nuclear accident at the Fukushima Daiichi nuclear power station, most NPPs have non-permanent (mobile) equipment to be used for accident mitigation, which may be outside the scope of safety classification. It is necessary to specify how non-permanent equipment, as well as radiation protection related equipment, emergency planning equipment and associated emergency response facilities, will either be included in the scope of an LTO programme or be on a separate, dedicated equipment list to be evaluated in terms of the respective safety factors in the PSR.

Consideration of safety significant SSCs identified by the probabilistic safety assessment (PSA) can also be incorporated into the scope setting process of LTO. This might involve identifying SSCs for normal operation, with no original safety function identified in the design, but credited in accidents or severe accident procedures. PSA could be used to determine component risk importance measures to provide a grading methodology for the in-scope equipment. Further information on the scope setting process for LTO can be found in Safety Reports Series No. 106, Ageing Management and Long Term Operation of Nuclear Power Plants: Data Management, Scope Setting, Plant Programmes and Documentation [5].

2.1.3. Scope of the most relevant safety factors for justification of LTO

The scope of the safety factors related to ageing needs to be expanded to include an evaluation of the time limited assumptions and assessments of ageing effects. In the review, increased importance is placed on ageing effects, degradation mechanisms, AMPs and obsolescence (see para. 4.6 of SSG-48 [2]).

Areas being evaluated in a PSR that provides justification for LTO are typically the same as in a normal PSR. The term 'safety factor', introduced in SSG-25 [1], refers to those aspects of safety of an operating NPP that are considered important and are not expected to be different during the period of LTO.

However, some of these important aspects of safety may gradually be challenged over time or extended operation because of physical degradation, ineffective knowledge and competence transfer to new plant personnel, evolution of technology or new developments in requirements, codes and standards. These aspects and their associated risks need to be effectively managed to ensure safe LTO. Within this document, the term 'relevant' is used for those PSR safety factors that relate to the analysis of these aspects of safety. A PSR that is used to support justification for LTO requires a more detailed analysis of the relevant safety factors, as they provide a demonstration of how the related risks are managed to ensure safe LTO.

Table 2 shows the most relevant safety factors required to support justification of LTO. The results of an in depth assessment of these safety factors provide important inputs for justification of LTO. The scope of SSCs is expanded to meet criteria according to SSG-48 [2], paras 5.16 and 5.17.

TABLE 2. THE MOST RELEVANT SAFETY FACTORS FOR JUSTIFICATION OF LONG TERM OPERATION *(according to SSG-48 [2])*

Safety factor	Name	Purpose of the review
SF1	Plant design	The review of this safety factor is crucial to the identification of additional safety improvements necessary to ensure that the licensing basis remains valid during the period of LTO. Such improvements might include refurbishment, the provision of additional SSCs, safety features and/or additional safety analysis and engineering justifications.
SF2	Actual condition of SSCs	The review of this safety factor determines the actual condition of in-scope SSCs and assesses whether they are adequate and capable of meeting design requirements at the end of LTO.
SF3	Equipment qualification	The review of this safety factor aims at determining whether plant equipment subject to qualification has been properly identified and qualified (including for environmental conditions) and whether equipment qualification status is preserved through an equipment qualification programme, maintenance, inspection and testing that provides confidence in the delivery of safety functions until at least the next PSR. The environmental qualification has been identified as a time limited ageing analysis (TLAA). TLAAs are re-evaluated for the planned period of LTO. The revalidation is to demonstrate that the equipment will maintain an adequate safety margin until the end of LTO.
SF4	Ageing	The review of this safety factor determines whether the ageing aspects affecting in-scope SSCs are effectively managed and whether an effective AMP is in place so that all required safety functions will be delivered until the end of LTO. The review of this safety factor identifies any plant programme enhancements needed to ensure that the structures or components will be able to perform their intended functions during LTO. The review of this safety factor also assesses whether the plant obsolescence management programme will remain effective for the period of LTO. TLAAs can be revalidated within this safety factor for the planned period of LTO. The revalidation demonstrates that the equipment will maintain its safety margin at the end of LTO.

2.1.4. Scope of other safety factors considered for justification of LTO

Additional areas within other PSR safety factors that can support justification of LTO are the capacity for long term storage of spent fuel and radiation waste, environmental monitoring and authorized discharges, plant safety performance and review of non-ageing related processes, such as the processes for corrective actions and operating experience.

There are several other safety factors whose areas of focus are also essential for justification of LTO. These safety factors, shown in Table 3, are typically used in synergy with the PSR. Where used to support LTO, they are to be thoroughly reviewed. The results of in depth assessment of these safety factors provide supplementary information needed for the justification of LTO.

TABLE 3. EXAMPLES OF THE USE OF OTHER SAFETY FACTORS TO SUPPORT JUSTIFICATION OF LONG TERM OPERATION

Safety factor	Name	Purpose of the review for LTO
SF5	Deterministic safety analysis	The review of this safety factor determines to what extent the existing DSA is complete and remains valid, taking into account the actual plant design, the actual condition of SSCs important to safety and their predicted state at the end of LTO, and the existence and adequacy of safety margins. An updated safety analysis report reflects the configuration of the plant that will operate during LTO. These updates include design changes such as replacements and upgrades of plant systems, new analyses and calculations using ageing related data, revalidation of TLAAs and other time limited assumptions (e.g. update of pressurized thermal shock analysis).
SF6	Probabilistic safety assessment	The review of this safety factor determines the extent to which the existing PSA study remains valid as a representative model of the NPP, whether it reflects the latest plant configuration and identifies weaknesses in the design and operation of the plant and whether it evaluates and compares proposed safety improvements in the global assessment. An adequate and up to date PSA model is an important precondition for appropriate selection of SSCs for the LTO scope setting.

TABLE 3. EXAMPLES OF THE USE OF OTHER SAFETY FACTORS TO SUPPORT JUSTIFICATION OF LONG TERM OPERATION (cont.)

Safety factor	Name	Purpose of the review for LTO
SF7	Hazard analysis	The review of this safety factor demonstrates the adequacy of protection against internal and external hazards, with account taken of the plant design, site characteristics, actual condition of the in-scope SSCs important to safety and their predicted state at the end of the LTO period. A comprehensive site reassessment could be required for justification of LTO.
SF8	Safety performance	The review of this safety factor determines whether the plant's safety performance indicators and records of operating experience, including the evaluation of root causes of plant events, are effective or indicate any need for safety improvements, and whether extrapolation of safety performance trends has been considered for the whole LTO period.
SF9	Use of experience from other plants and research findings	The review of this safety factor determines whether there is adequate feedback of relevant experience from other NPPs and research from the findings, and whether this is used to introduce reasonable and practicable safety improvements at the plant or operating organization. Consideration of the latest international experience and research findings related to LTO could be a special focus for the review of this safety factor when the PSR is used to support LTO.
SF10	Organization, management system and safety culture	The review of this safety factor determines whether the organization, the management system and the safety culture are adequate and effective to ensure the safe operation of the NPP. This safety factor is an important precondition for safe LTO. The review determines whether an adequate policy regarding LTO is present and whether the effectiveness of dedicated organizational structures and sufficient resources will be warranted for the duration of LTO.

TABLE 3. EXAMPLES OF THE USE OF OTHER SAFETY FACTORS TO
SUPPORT JUSTIFICATION OF LONG TERM OPERATION (cont.)

Safety factor	Name	Purpose of the review for LTO
SF11	Procedures	The review of this safety factor determines whether the operating organization's processes for managing, implementing and adhering to operating and working procedures, as well as for maintaining compliance with operational limits, conditions and regulatory requirements, are adequate and effective and ensure plant safety for the period of LTO.
SF12	Human factors	The review of this safety factor evaluates the various human factors that may affect the safe operation of the NPP and seeks improvements that are reasonable and practicable for the whole period of LTO. Issues related to the availability of sufficiently qualified staff, including the effective knowledge and competence management necessary for the LTO period, are included in the review of this safety factor.
SF13	Emergency planning	The review of this safety factor determines whether the operating organization has in place adequate plans, staff, facilities and equipment for dealing with emergencies; and whether the operating organization's arrangements have been adequately coordinated with the arrangements of local and national authorities and are regularly exercised. Due consideration of changes at the plant site, its surroundings and the status of equipment and facilities used for emergency preparedness are provided by the review of this safety factor to confirm their pertaining adequacy during LTO.
SF14	Radiological impact on the environment	The review of this safety factor determines if the operating organization has an adequate and effective programme for monitoring the radiological impact of the plant on the environment, which ensures that emissions are properly controlled and are as low as reasonably achievable. This review includes the impact on the environment of the activities carried out during the LTO period of the plant (e.g. refurbishment, replacement, new waste storage).

2.2. PERIODIC SAFETY REVIEW BASIS DOCUMENT

The PSR basis document defined in para. 4.6 of SSG-25 [1] is a key document in the PSR. It defines the scope of the review and how it will be conducted. It serves as the agreement between the utility and the regulator regarding the PSR's scope and methodology.

The intent of SSG-25 [1] is not specifically to provide recommendations for the activities performed during LTO of an NPP. However, the scope of the review of the safety factors, which is specified in the PSR basis document, can be adapted to support an assessment of the feasibility of LTO, as indicated in para. 2.10 of SSG-25 [1].

2.2.1. Development of the PSR basis document

The PSR basis document describes how the review of safety factors is conducted to meet the key objectives of the PSR. The PSR basis document is developed taking into account how it fits into the overall LTO safety justification. When defining the scope of the PSR, it is important to consider the stage of the justification of LTO at which the PSR is being conducted.

The responsibility for developing the PSR basis document rests with the licensee, in accordance with the requirements and guidance issued by the regulatory body (see Section 5 for details on roles and responsibilities). The quality and completeness of the PSR basis document significantly affect the efficiency of performing a PSR, the quality of the results and eventually the success of the PSR in meeting its objectives. An independent review of the PSR basis document may be required to ensure the expected results are obtained, especially in the case of a first PSR or a major update of a previously used PSR basis document. This major update could include changes incorporated into the PSR basis document in support of justification of LTO. Independent review by the IAEA or other recognized international organizations could be considered.

In order to facilitate the production of the PSR basis document and obtain the regulatory body's concurrence, it is pragmatic to have different cut-off dates: one for the codes and standards against which the review will be conducted and another, later one for the NPP status. This will facilitate the identification of, and agreement on, applicable requirements and documents that will define the scope of the review. Appropriate cut-off dates are determined in accordance with the adopted strategy and time schedules for PSR and LTO, included in the PSR basis document and agreed with the regulatory body.

When establishing the project schedule for the PSR, it is important to consider the regulatory requirements regarding the submission date for both the PSR and the justification of LTO. This is to ensure a suitable time frame to support justification of LTO, as well as adequate time necessary for the resolution of any deviation identified in the PSR that could hamper LTO.

2.2.2. Content and structure of the PSR basis document

Appendix II of SSG-25 [1] provides the recommended content of the PSR basis document. The proposed content and the structure of this PSR basis document example are also suitable for a PSR that will be used to support LTO.

If the PSR is performed concurrently with the preparation for LTO, the PSR basis document typically includes a description of interactions between the PSR and LTO projects. This could include the communication strategy and interface meetings between the two teams to take advantage of any potential synergies, as well as meetings with the regulatory body involving both teams.

A global assessment is a part of the PSR basis document (see section 6 of SSG-25 [1]). This assessment provides the methodology for processing the results of the safety factor review with the objective of defining an integrated implementation plan and providing justification for LTO. The procedures of the global assessment are generally robust and generic in order to successfully accommodate the LTO assessment and its results, but the assessment period is to be extended to the period of LTO. Based on the expected schedules of the PSR and LTO projects and their anticipated synergies, the global assessment could be reviewed and updated as necessary (see Section 4).

2.2.3. Regulations, codes and standards considered in the PSR

The PSR basis document contains a list of applicable regulations, codes and standards, as well as clear criteria for selecting those that will be used in the PSR. The list could be updated to reflect the date of the review. It typically includes:

(a)   National laws and regulations;
(b)   Operation licence and other relevant licences;
(c)   IAEA Safety Standards;
(d)   Codes and standards issued by the State of the vendor of the technology;
(e)   The latest revisions of these codes and standards.

Other requirements, such as international safety standards and operating practices, as well as national or international guides, are to be met as much as practicable. The selection and hierarchy of safety standards and operating practices considered are clearly stated in the PSR basis document, with special consideration given to safety standards issued by the State of the vendor of the technology.

Reference can also be made to international codes and standards (such as those of the IAEA, the International Organization for Standardization (ISO) or the International Electrotechnical Commission (IEC)) or, where appropriate, to the codes and standards of a recognized organization of a particular State (for example, the American Society of Mechanical Engineers (ASME), the Nuclear Safety Standards Commission (Kerntechnischer Ausschuss) or the Institute of Electrical and Electronics Engineers (IEEE)).

The practices of international organizations, such as good practices collected by the World Association of Nuclear Operators (WANO) and the IAEA, as well as the information generated by owners' groups, could also be relevant and taken into account.

2.3. METHODOLOGY AND ACCEPTANCE CRITERIA

2.3.1. Introduction

This section describes the methodology and criteria entailed in a review of safety factors in PSR, focusing on the most relevant ones for justification of LTO. Suggested adjustments to the scope and content are proposed and discussed where the PSR is intended to support an LTO assessment.

The main safety aspects to be evaluated in a PSR are typically grouped into five categories, which pertain to the plant (safety factors 1–4), safety analyses (safety factors 5–7), safety performance and feedback from operating experience (safety factors 8–9), management (safety factors 10–13) and the environment (safety factor 14).

The safety factors most related to LTO are safety factor 1 (plant design), safety factor 2 (actual condition of SSCs), safety factor 3 (equipment qualification) and safety factor 4 (ageing) (see Table 2). Other safety factors, as shown in Table 3, are also relevant for LTO.

2.3.2. Review of safety factors

General review recommendations for the PSR safety factors are provided in paras 4.18–4.20 and paras 5.4–5.14 of SSG-25 [1]. The methodology and the criteria for the review of safety factors, agreed upon with the national regulatory body, are provided in the PSR basis document. The review of the safety factors aims to determine the following:

(a)   The degree of validity of the licensing basis and safety documentation;

(b)   The adequacy and effectiveness of provisions that ensure the safety of the installation until the next PSR, or if the PSR is used for supporting justification of LTO, for the whole period of LTO;

(c)   The extent to which the plant complies with current national and international regulations, guidelines and good practices in the field of nuclear safety;

(d)   The adequacy of plant programmes considered in the safety factors relating to plant design, actual condition of in-scope SSCs, equipment qualification and ageing;

(e)   If the management system adequately addresses quality management and configuration management;

(f)   If TLAAs are valid for the period of LTO;

(g)   If programmes for promoting safety culture focus on the pursuit of excellence in all aspects of safety management and human factors.

The review of the safety factors is an iterative process that requires coordination and integration among the teams carrying out the evaluation of the individual safety factors. Although the methodology of the review differs among safety factors, the evaluation focuses on identifying positive and negative findings with respect to the scope of each safety factor. Findings from the safety factor reviews are to be further assessed in the PSR global assessment with respect to their safety relevance. Findings from different international peer review services may provide valuable inputs into the evaluation of certain safety factors or may support the evaluation. Examples include IAEA technical safety review (TSR) services[2], safety aspects of LTO (SALTO)[3] peer review service for safety factors relevant to ageing management and LTO, and operational safety review team (OSART)[4] peer reviews for assessing operational aspects of the plant.

To obtain structured and comparable results, harmonization of methodical approaches for reviewing each safety factor is essential. This can be achieved by setting out each safety factor into a clearly defined structure, typically based on regulatory requirements or plant operating practices. Most of the activities carried out at an NPP are managed and organized by programmes and processes with defined content, methodologies with defined review criteria and quality management systems, as well as interfaces with each other. The established

---

[2]   IAEA technical safety review services include probabilistic safety assessment review, design safety review, generic reactor safety review, PSR and accident management review.

[3]   An IAEA peer review service focusing on ageing management and organizational aspects of long term operation. A high level review of the PSR programme may be included in the scope.

[4]   An IAEA peer review service focusing on operational safety. It includes the review of ageing related aspects, as well as programmatic review of PSR.

structure of an individual safety factor could thus, as far as possible, support a review based on a 'process' approach, consisting of two levels as follows:

(a) High level review, where the review focuses on whether documentation for management and processes, high level policies and methodologies set adequate and clear requirements for fulfilling specific criteria.

(b) Detailed review, which focuses on a particular situation at the plant addressed and managed by one of the aforementioned high level documents or processes; it is verified whether the established requirements are fulfilled and implemented in practice. Evidence of verification may be based upon procedures, records, schemes, measurements and tests.

2.3.2.1. *Applicable criteria*

Selected criteria for review are typically brief, unambiguously formulated statements that provide a good basis for verifying the conformity of a specific aspect of a PSR's safety factor. The basis for the formulation of a criterion is provided by referencing documents, whose order of priority is determined by their legally binding nature. Possible sources of reference documentation can be, among others, national laws and regulations, regulatory guidance and requirements, technical standards, IAEA requirements and safety guides or WANO documents. Where applicable, Western European Nuclear Regulators' Association (WENRA) safety reference levels and documents of reviews organized by the European Nuclear Safety Regulators Group (ENSREG) may also serve as reference. Experience, good technical practices and findings from research and development can provide inputs as to which criteria can be considered. In some cases, criteria can be identified by means of surveys, interviews and analysis of audits.

The applicable criteria for reviewing the safety factors are established prior to the review of safety factors in the PSR. These are discussed and agreed with the national regulatory body and thus set the desired level of review detail. This leads to a variable level of detail of the review (see para. 5.5 of SSG-25 [1]) and to different review strategies tailored to the national regulatory process (see para. 4.4 of SSG-25 [1]). For example, some safety factors might be evaluated at the higher system level, rather than at the level of components. An LTO assessment requires a systematic, per component analysis, as described in section 7 of SSG-48 [2]. This difference needs to be considered when using a PSR for justification of LTO, and the level of detail of a PSR needs to be adjusted to fit the scope of LTO. If the PSR is to be used in support of an LTO assessment, a relevant underlying methodology could be to follow SSG-25 [1], where several criteria are provided for reviewing each safety factor in the PSR.

Sections 2.3.2.2–2.3.2.5 below describe the methodology for reviewing the plant related safety factors when considering LTO, including some adjustment to the individual safety factor's methodology in cases where the PSR is used to support LTO.

2.3.2.2. *Safety factor 1 (plant design)*

The methodology for reviewing this safety factor is described in paras 5.19–5.26 of SSG-25 [1]. Examples of aspects of an LTO programme covered by safety factor 1 include, among others, a review of the list of SSCs important to safety, including its documentation, and the compliance of the SSCs' actual status with the plant design (configuration management). Continuous plant processes in scope for this safety factor, such as the licensing basis management programme, the final safety analysis report (FSAR) management programme and configuration management, can also be used in an LTO programme. In cases where the PSR is used in support of LTO, the list of in-scope SSCs and out of scope SSCs specified in paras 5.16 and 5.17 of SSG-48 [2] is used. This list and other relevant documentation are used to establish the compliance of current design basis with current nuclear safety standards, as well as to identify differences. Where the PSR is to be used in support of LTO, any necessary safety improvements are identified (see para. 3.5 of SSG-25 [1]). Benchmarking the design against other similar installations of comparable age can identify possible modifications to improve plant safety. In the specific case of a PSR supporting LTO, such a benchmarking could focus on NPPs that have already undergone an LTO assessment. In the review of safety factor 1, according to para. 5.19 of SSG-25 [1], a clause by clause review of the listed standards is to be performed.

2.3.2.3. *Safety factor 2 (actual condition of SSCs important to safety)*

The methodology for reviewing this safety factor is described in paras 5.30–5.36 of SSG-25 [1]. Examples of aspects of an LTO programme covered by safety factor 2 include, among others, the list of in-scope SSCs, their classification and intended function, a review of the in-scope SSCs' functional capability, a review of in-service inspection and maintenance programmes, and the assessment of the current physical status of in-scope SSCs. The current design basis and the list of in-scope SSCs, evaluated as part of safety factor 1, are required for the review of this safety factor. Plant programmes (maintenance, in-service inspections, surveillance and monitoring) are checked against the nine attributes of an effective AMP provided in table 2 of SSG-48 [2].

One of the preconditions to LTO is that documentation relevant for the LTO assessment, such as plant programmes, is in place at the plant (see para.

4.1 of SSG-48 [2]). Plant programmes are in scope for safety factor 2, meaning that the review of this safety factor has to be performed for the verification of preconditions when the PSR is used to support LTO.

2.3.2.4. *Safety factor 3 (equipment qualification)*

The methodology for reviewing this safety factor is described in paras 5.42–5.44 of SSG-25 [1]. Examples of aspects in an LTO programme covered by safety factor 3 include, but are not limited to, the review of the qualification status of in-scope SSCs, review of the equipment qualification programme and an evaluation of activities for preserving the equipment qualification (e.g. environmental monitoring). Continuous plant processes in scope for this safety factor that can be used for justification of LTO are programmes for the plant's lifetime management, the proactive obsolescence programme and programmes for the replacement of major components. The list of in-scope SSCs, evaluated as part of safety factor 1, is required for the review of this safety factor. The adequacy of the organization's equipment qualification programme is assessed by reviewing the available documentation and the environmental conditions of qualified SSCs, checking qualification databases for completeness and assessing the effectiveness of the equipment qualification programme by means of plant walkdowns. One of the preconditions to LTO is that an equipment qualification programme is in place at the plant (see Section 3.1.2 of this publication and para. 4.1 of SSG-48 [2]).

2.3.2.5. *Safety factor 4 (ageing)*

The methodology for reviewing this safety factor is described in paras 5.49–5.51 of SSG-25 [1]. Examples of aspects of an LTO programme covered by PSR relating to safety factor 4 include, but are not limited to, review of the ageing management strategy, review of AMPs and revalidation of TLAAs. The review of safety factor 4 ensures that the ageing effects are adequately managed for the next PSR for the duration of LTO (see para. 5.51 of SSG-25 [1]). If the PSR is used to support justification of LTO, the review of the ageing management strategy of the organization has to take into account a systematic approach described in SSG-48 [2], summarized in the plan–do–check–act cycle (also known as the Deming cycle).

A strategy for addressing technological obsolescence is also part of this safety factor's review (see para. 5.48 of SSG-25 [1] and section 6 of SSG-48 [2]). In-scope SSCs are typically clustered into commodity groups in order to perform the AMR (see para. 5.20 of SSG-25 [1]). The use of commodity groups could thus be considered when reviewing safety factor 4 if the PSR is used in support of LTO. The scope of this safety factor needs to be expanded by including the re-evaluation of TLAAs (see para. 3.6 of SSG-25 [1]), such as radiation embrittlement of the reactor vessel and fatigue analysis. For an extended list of TLAAs, see appendix II of Safety Reports Series No. 82 (Rev. 1), Ageing Management for Nuclear Power Plants: International Generic Ageing Lessons Learned (IGALL) [6].

3. COORDINATION OF PERIODIC SAFETY REVIEW AND LONG TERM OPERATION ACTIVITIES

3.1. AREAS IN WHICH THE LONG TERM OPERATION PROGRAMME COULD PROVIDE INPUTS TO PERIODIC SAFETY REVIEW

This section illustrates how PSR and LTO programmes can be coordinated to avoid repeating, recurrent or overlapping activities and to ensure that no important activities are omitted.

In areas where significant work has already been completed in the programme for LTO, the PSR of those areas could be limited to programmatic reviews. Those safety factors where this opportunity exists include safety factor 2 (actual condition of SSCs important to safety), safety factor 3 (equipment qualification) and safety factor 4 (ageing). The selected programmatic approach to incorporate assessment results from LTO into a PSR needs to be reflected in the PSR basis document with an appropriate justification (see para. 5.5 of SSG-25 [1]), as well as agreed with the regulatory body.

This section describes potential inputs expected from the LTO programme to the PSR programme and the stage at which the inputs could be ideally obtained. This section also provides considerations on the sharing of responsibilities and coordination between the LTO programme and the PSR programme necessary to justify LTO.

3.1.1. Actual condition of SSCs important to safety (safety factor 2)

When the timing of a PSR for the plant comes close to the intended period of LTO, certain activities inherently required for the development of a programme for LTO, as described in SSG-48 [2], could be used to support the review of safety factor 2. The methodologies for the review of safety factors presented in SSG-25 [1], for which synergies exist within the guidance in SSG-48 [2], include the following:

(a) The current condition of SSCs (see para. 5.30 of SSG-25 [1]). This is a review of the current condition of the plant using knowledge of relevant ageing effects and non-physical (obsolescence) issues. The review considers the expected condition of the plant over the PSR period or the intended period of LTO if the PSR is being used directly to justify LTO.

(b) The adequacy of plant programmes that support confidence in the condition of SSCs (see paras 5.31 and 5.32 of SSG-25 [1]). This consists of a review of the plant programmes (e.g. maintenance, obsolescence) and programmes that are used to demonstrate the condition of SSCs (e.g. in-service inspection, surveillance, water chemistry).

(c) The alignment between the current condition of SSCs and their design basis (see paras 5.35 and 5.36 of SSG-25 [1]). This review assesses the implications of changes to design requirements or standards. In addition, the current and anticipated condition of the plant is checked for consistency against the current design basis.

Major changes in the plant's operational strategy (e.g. cycle length, flexible generation) or power rating could challenge the effectiveness of using the current condition of SSCs in assessments of future performance. These changes may coincide with a plant's LTO as commercial considerations change or major enhancements are implemented.

Sections 3.1.1.1 and 3.1.1.2 below identify those key steps in developing a programme for LTO that could directly support an assessment of safety factor 2 during the PSR. The extent to which LTO activities described in these sections may be beneficial to the PSR depends upon their timing and a mutual understanding of the requirements, processes and constraints associated with the two programmes of work. Identification of synergies at an early stage of the LTO programme is likely to offer significant and beneficial input into the review of safety factor 2. Although this section focuses on synergies between LTO programme activities and safety factor 2, it is recognized that other safety factors, notably safety factor 3 (equipment qualification), safety factor 4 (ageing), safety factor 5 (deterministic safety analysis) and safety factor 7 (hazard analysis), can also benefit from the output of the programme for LTO in this area.

3.1.1.1. *Preconditions for LTO assessments*

Where existing plant programmes are to be credited for managing ageing during LTO, they need to be comprehensive and properly implemented.

Programmes monitoring the performance of, or detecting ageing effects of, in-scope SSCs [2] are particularly relevant for safety factor 2 and include the following:

(a) Maintenance;
(b) In-service inspection;
(c) Surveillance;
(d) Water chemistry.

Data from these activities, from the plant's maintenance programme and from targeted walkdowns, are used for trending and evaluating the conditions of in-scope SSCs. These data are retained in an auditable format. These programmes need to be consistent with the nine generic attributes of an effective AMP as defined in table 2 of SSG-48 [2]. Data presented in the System Health Report, or other records credited for the management of ageing, also need to be checked for consistency with the observed condition of in-scope SSCs, as well as for their effectiveness in assessing and mitigating ageing degradation.

3.1.1.2. *AMR for LTO*

A key activity of an LTO programme is an AMR, which is a systematic review of the ageing management of in-scope SSCs [2] that will, for each SSC, identify and assess ageing effects and degradation mechanisms, both potential and already observed. This review will identify the AMPs, new and existing, applicable to the SSCs whose effectiveness is assessed against the nine generic attributes of an AMP described in table 2 of SSG-48 [2]. The AMR for LTO consists of the following steps:

(a) Assessment of the current condition of the structure or component;
(b) Identification of ageing effects and degradation mechanisms based on fundamental knowledge for understanding ageing;
(c) Identification of the appropriate programme for ageing management;
(d) Reporting on the AMR.

It is necessary to evaluate the impact of ageing on the capability of SSCs to perform their intended safety functions, so the current condition of those plant items is appraised. According to para. 5.32 of SSG-25 [1], "[t]he validity of existing records should be checked to ensure that they accurately represent the actual condition of the SSCs important to safety, including any significant findings from ongoing maintenance, tests and inspections."

In order to ensure that the scope of the credited AMP is comprehensive, an evaluation of the condition of SSCs should include degradation mechanisms that have the potential to affect safety performance, the current and postulated service and environmental conditions, the range of installation sites and installation or maintenance history.

Based upon a fundamental understanding of the SSCs' safety functions, current condition and the relevant ageing degradation mechanisms, the AMR will identify the required ageing management strategy. Existing AMPs can then be reviewed for applicability to the SSCs under consideration and their completeness. The need for new AMPs or supplementary activities may be identified in areas where critical data are identified as missing, in order to establish adequate ageing management. Such activities could include the following:

(a) Grouping SSCs important to safety for the purpose of AMR according to functional system or type, as recommended in para. 5.34 of SSG-25 [1].

(b) Performing special tests or monitoring the in-scope SSCs and the stressors responsible for ageing. Paragraph 5.32 of SSG-25 [1] states: "Where data are lacking, they should be generated or derived by performing special tests, plant walkdowns and inspections as necessary."

(c) Performing and evaluating research and development activities.

(d) Replacing the in-scope SSC with one for which ageing is successfully managed.

(e) Reviewing the established good practices in ageing management of in-scope SSCs that are comparable in terms of their equipment:
   — Function (performance requirements under normal and design basis accident conditions, noting the qualified condition and/or post fault claim);
   — Materials;
   — Form (technology and physical construction);
   — In-service condition (environmental, process or operating regime).

(f) Reducing uncertainties. Paragraph 5.33 of SSG-25 [1] states that "uncertainties may be reduced by considering evidence from similar components from other plants or facilities that are subject to similar conditions and/or knowledge of the relevant ageing processes and operating conditions."

The AMR will result in the identification of one or more AMPs for each in-scope SSC. The AMPs will either meet the nine attributes of an effective AMP or corrective actions will need to be identified and managed to ensure that comprehensive AMPs will be established. The status of these programmes, including the maintenance and up to date status of records, are part of the consideration of the actual condition of SSCs and are key inputs for PSR safety factor 2.

Safety factor 2 can benefit from following the guidance for the development of a programme for LTO provided in paras 7.1 to 7.41 of SSG-48 [2], meeting the preconditions for LTO, completing the scoping process and establishing effective AMPs resulting from a comprehensive AMR.

3.1.2. Equipment qualification (safety factor 3)

3.1.2.1. Concept and process for equipment qualification

Paragraph 4.48 of SSR-2/2 (Rev. 1) [3] states: "Appropriate concepts and the scope and process of equipment qualification shall be established, and effective and practicable methods shall be used to upgrade and preserve equipment qualification."

The equipment qualification is required to demonstrate that the equipment will be capable of performing its intended safety function(s) under the range of service conditions specified for the nuclear installation in operational states and in accident conditions. This includes an evaluation of the ability of systems or components to perform these safety functions under the effects of specified service conditions during plant states and during external events included in the design of the nuclear installation (e.g. seismic events).

In contrast, internal fires, explosions, internal flooding, tornadoes and hurricanes are not normally considered in the equipment qualification programmes because the design generally protects the equipment from the effects of these events. One of the objectives of equipment qualification is the prevention of common cause failures arising from the exposure of equipment to the specified service conditions.

The equipment qualification process comprises three phases:

(a) Establishment of appropriate design inputs (e.g. safety functions, service conditions);
(b) Establishment of equipment qualification process steps (e.g. type testing, analysis, operating experience, combined methods);
(c) Preservation of the status of qualified equipment.

The equipment qualification includes seismic qualification, environmental qualification for mild and harsh environments (e.g. temperature, pressure, humidity, contact with chemicals, radiation exposure, meteorological conditions, submergence and ageing mechanisms), and electromagnetic qualification (e.g. effects of electromagnetic or radiofrequency interference).

Although seismic qualification generally applies to all SSCs, the environmental qualification applies primarily to electrical and active mechanical equipment, instrumentation and control (I&C) equipment, and components associated with this equipment (e.g. seals, gaskets, lubricants, cables, connections, mounting and anchoring structures).

The qualification process for passive mechanical components (e.g. piping and vessels), for which the safety performance is assured by design, is performed in accordance with applicable codes.

3.1.2.2. *Review of equipment qualification*

The objective of the review of equipment qualification within safety factor 3 is to determine whether plant equipment subject to qualification has been properly qualified (including for environmental conditions) and whether this qualification, as discussed above, is being preserved through an adequate programme of maintenance, inspection and testing that provides confidence in the delivery of safety functions until at least the next PSR (see para. 5.38 of SSG-25 [1]). The review of safety factor 3 typically includes an assessment of the following:

(a) Changes in the equipment classification resulting from design modifications;
(b) Qualification for all designed environmental conditions;
(c) The availability of equipment that is required to fulfil safety functions;
(d) Confirmation of the state of the equipment and any necessary supporting structures (such as equipment anchorage for seismic) that are necessary to ensure the equipment qualification;
(e) Quality management provisions that ensure that an effective qualification programme is in place.

Walkdowns and inspections are carried out to verify that the installed equipment matches the required qualification described in the safety documentation and to provide an input into the review of the adequacy of the plant's procedures for maintaining equipment qualification.

The plant equipment (or its component parts) that is subject to environmental qualification has different time frames for which the qualified life is established (e.g. a qualified sensor has a shorter qualified life, whereas a qualified cable has a longer qualified life). Before the equipment (or some of its parts) can reach the end of its qualified life, it is either replaced at time specified intervals (e.g. through an equipment replacement programme) or reassessed during its lifetime, taking into account the actual operating environment of the equipment and current understanding of equipment degradation mechanisms. In this regard, the PSR needs to verify whether equipment whose qualified life is about to expire will be replaced in accordance with a time specified interval, or whether the equipment can be requalified so that it can maintain its qualification status for the period until the next PSR.

3.1.2.3. *Preservation of equipment qualification — a precondition for LTO*

A preservation of equipment qualification is required for the entire lifetime of the NPP. Preservation of equipment qualification is an integral part of the equipment qualification programme, which is a precondition for LTO. The preservation activities that are performed during the entire plant lifetime include the following:

(a) A periodic surveillance of qualified equipment, which ensures that:
   — The operation and maintenance activities do not compromise the status of qualified equipment by changing the configuration, mounting orientation (horizontal or vertical supports) or electrical, pneumatic or hydraulic interfaces;
   — The systems and components continue to meet their performance requirements;
   — Abnormalities in the configuration of the equipment are detected and corrective actions are completed in a timely manner to preserve the status of qualified equipment;
   — The criteria for identifying premature ageing degradation are specified;
   — During the periodic surveillance, if unexpected degradation is observed, the effect of this degradation on the capability of the equipment to perform the assigned safety function is evaluated.
(b) A periodic or continuous monitoring of environmental conditions in both mild and harsh environmental zones to ensure that the assumptions used to determine the equipment's qualified life remain unchanged. For example, normal operating temperature in the reactor building is lower than considered in the design of environmentally qualified components (except core neutron detectors). In this regard, the original time limited assumptions related to ambient temperature in the reactor building remain valid.
(c) Detection of any hot spots (e.g. elevated temperatures, elevated radiation, humidity, chemical impact) that could result in a significant degradation mechanism originally not considered when establishing the equipment qualified life.
(d) Periodic replacement of component parts at time specified intervals (e.g. seals, gaskets, lubricants, filters) that are more susceptible to degradation. Such parts may need to be periodically replaced (i.e. as opposed

to being reused) during maintenance activities specifically undertaken for equipment qualification purposes.

(e) Monitoring the storage conditions of the qualified equipment and spare parts.

(f) Planning for obsolescence (e.g. records of the non-availability of replacement components from the original equipment manufacturer and the acceptability of appropriately qualified replacement components).

(g) Configuration management (change control) to ensure that the implications of equipment qualification are appropriately considered whenever changes occur to the installation, to equipment or to operating, maintenance or replacement activities.

3.1.2.4. *Inputs from the LTO programme to the assessment of safety factor 3 in the PSR*

An LTO programme includes a comprehensive review of the equipment qualification programme. The equipment subject to qualification is identified and included in the scope for LTO assessment consistently with an approach on scope setting for SSCs for LTO, as defined in paras 5.14–5.21 of SSG-48 [2].

The scope of review of equipment qualification for LTO is comparable to a standard review of safety factor 3 in the programme for PSR. However, the results of the re-evaluation of the TLAA related to equipment qualification may impact the initial assumptions (e.g. seismic, environmental parameters, electromagnetic interference) used for establishing the equipment qualification. In this case, the equipment qualification programme needs to be reviewed for potential impact. The review of equipment qualification for LTO includes the results of a comprehensive review of the following subjects:

(a) Monitoring of environmental conditions to verify whether:
— The assumptions in the equipment qualification are consistent with the ambient conditions in the part of the installation in which the equipment is installed;
— The design limits of the equipment are not being exceeded;
— The status of qualified equipment remains valid.

(b) Monitoring of the condition of qualified equipment to investigate whether:
— The initial assumptions used to establish an ageing programme are still valid;
— There are additional ageing effects not identified when the equipment qualification was initially established.

28

(c)   Evaluation of ageing effects on the qualified status of the equipment to verify whether:

— Any new ageing effect or increase in the effects of previously known ageing effects are identified that would require the equipment qualification programme to be reviewed to determine whether changes in the qualified life or maintenance of the equipment are needed;
— Periodic preventive maintenance, predictive maintenance, equipment calibration, surveillance, testing, condition monitoring, corrective action, identification of trends in equipment failures and operating experience reviews allow for identifying and mitigating unanticipated ageing degradation that was not accounted for when establishing the original equipment qualification;
— Appropriate condition indicators for a given type of equipment are selected to help detect changes caused by significant ageing effects.

(d)   Revalidation of the TLAA related to equipment qualification.

The results of processes that identify ageing related failures or significant material degradation of qualified equipment are used to assess the need to revise the maintenance, surveillance and replacement programmes that are related to equipment qualification.

3.1.2.5. *Revalidation of TLAAs*

As part of the revalidation of TLAAs, the environmental qualification is reassessed to confirm that time limited assumptions made when establishing the equipment's qualified life remain unchanged, or, if these were changed, to establish how this affects the plant equipment qualification programme and update it accordingly. The revalidation of environmental qualification is performed when the intended period of plant operation is extended to demonstrate that the equipment will maintain its safety margin until the end of the extended plant life. Revalidating TLAAs during an LTO assessment could provide useful inputs into safety factor 3 where the effects of ageing degradation (e.g. due to irradiation, corrosion, humidity) are taken into account for equipment qualification (see paras 5.39 and 5.40 of SSG-25 [1]). The methodology for revalidating equipment qualification consists of the methods and criteria for confirming that the initial assumptions used for environmental qualification are either valid or have changed. Revalidation of TLAA confirms whether:

(a)   The environmental zones and associated environmental parameters have changed;

(b) The equipment installed within the environmental zones has adequate qualification;

(c) The effects of ageing on the intended function(s) remain valid for the duration of LTO;

(d) Any new significant degradation mechanism that could impact the status of the qualified equipment is identified;

(e) Activities prescribed to preserve the qualification are still comprehensive.

If the revalidation of TLAAs shows that the initial assumptions used for establishing the equipment's qualified life have changed, the equipment qualification programme needs to be reviewed and corrective actions defined and implemented in a timely manner. Examples of corrective actions include the following:

(a) Revision of the equipment qualification master list (e.g. adding new equipment for which significant ageing degradation mechanisms were identified);

(b) Reassessment or requalification of equipment (or qualification of the newly installed equipment) for the new environmental parameters (e.g. seismic, environmental parameters due to accident conditions, electromagnetic interference, radiofrequency interference);

(c) Replacement of equipment that did not pass requalification or for which the requalification could not confirm a qualified life for the entirety of the extended plant life;

(d) Modification of interfacing plant programmes (e.g. operation, maintenance, ageing management).

A methodology to extend the equipment qualified life (e.g. cables) may consider reducing the excessive conservatism incorporated into the original qualification basis. For example, the excess conservatism adopted by the ageing model (e.g. Arrhenius model) to extend the qualified life may be reduced by:

(a) Evaluating the initial assumption of the actual ambient and/or operating temperature;

(b) Lowering overly conservative activation energy, as discussed in IGALL [6] (and similarly in section X.E1, Environmental Qualification of Electric Equipment, of NUREG-2191, Vol. 2, Generic Ageing Lessons Learned for Subsequent License Renewal (GALL-SLR) Report [7]).

For example, the reanalysis could replace the initially assumed ambient temperature in the Arrhenius model that was used to derive the ageing temperature and duration. If the component was operating at a lower ambient and/or operating temperature, the reanalysis would result in a longer qualified life. The opposite is also true for higher actual ambient and/or operating temperatures.

3.1.2.6. Reassessment of the qualified life of equipment

The qualified life of equipment is reassessed throughout the lifetime of the installation to take into account changes in the actual service conditions, such as temperature and radiation levels, and development in the knowledge and understanding of degradation mechanisms. Consequently, an evaluation of the data from monitoring environmental conditions and the condition of equipment is typically used to reassess the qualified life of equipment. A well established and documented technical basis is necessary to support an extension of the qualified life of equipment. In addition, any conclusions regarding the status of qualified equipment are re-evaluated to consider any changes in performance requirements or plant conditions. For example, methods such as re-evaluation of the conservativism of assumptions made in the original equipment qualification, type testing of naturally aged equipment with additional ageing to support the extension of the qualified life, and equipment replacement and refurbishment are used for reassessing the qualified life. Changes in the stressor intensity (e.g. changes in temperature and radiation levels) may also be evaluated to reassess the qualified life.

3.1.2.7. Review of the equipment qualification programme

The equipment qualification programme is a relevant programme to LTO. Crediting the equipment qualification programme for LTO includes evaluating its consistency with the nine attributes of an effective AMP, provided in table 2 of SSG-48 [2], to determine whether the programme is effective in detecting, monitoring and preventing or mitigating any effects and degradation mechanisms that could impair the qualification status of the components. If the evaluations conclude that the existing equipment qualification programme is not sufficiently effective, this programme needs to be improved or modified, consistent with the nine attributes in table 2 of SSG-48 [2].

3.1.2.8. IAEA framework for equipment qualification

IAEA Safety Standards Series No. SSG-69, Equipment Qualification for Nuclear Installations [8], provides guidance on how to meet the relevant safety requirements established in SSR-2/1 (Rev. 1) [9] and SSR-2/2 (Rev. 1) [3]. Specifically, SSG-69 [8] provides recommendations on a structured approach for establishing and preserving equipment qualification in nuclear installations, in order to minimize exposure to common cause failures of the equipment. Qualification methods described in SSG-69 [8] apply to both nuclear grade products and commercial grade products that need to be assessed or qualified to confirm their suitability to meet the functional and performance requirements while operating within specified service conditions.

3.1.3. Ageing (safety factor 4)

In-scope SSCs are subject to some form of physical change caused by ageing, which could eventually impair their safety functions and service lives. To control the ageing degradation of SSCs within acceptable limits, an AMP for the NPP is implemented that covers the systematic arrangements aimed at providing adequate ageing management measures, specifically engineering, operations and maintenance actions. The basic concept of ageing management as presented in SSG-48 [2] covers both physical ageing and non-physical ageing.

Physical ageing represents ageing of SSCs due to physical, chemical and/or biological processes. Physical ageing is managed using AMPs and other plant programmes (further discussed in Section 3.1.3.1) and using activities resulting from TLAA revalidation (further discussed in Section 3.1.3.2).

Non-physical ageing is referred to as 'obsolescence'. There are three different types of obsolescence (for details, see also table 1 in SSG-48 [2]): (i) obsolescence of technology; (ii) obsolescence of regulations, codes and standards; and (iii) obsolescence of knowledge. Technological obsolescence is discussed further in Section 3.1.3.3.

Within the PSR, the cumulative effects of ageing on NPP safety, the effectiveness of AMPs and the need for improvements to AMPs, as well as technological obsolescence, are all covered and reviewed in safety factor 4 (ageing).

3.1.3.1. Review and improvement of plant programmes and AMPs for LTO

The main IAEA requirements relevant to AMPs and LTO are as follows:

(a) Paragraph 4.50 of SSR-2/2 (Rev. 1) [3] states:

"The ageing management programme shall determine the consequences of ageing and the activities necessary to maintain the operability and reliability of structures, systems and components. The ageing management programme shall be coordinated with, and be consistent with, other relevant programmes, including the programme for periodic safety review."

(b)   Paragraph 4.53 of SSR-2/2 (Rev. 1) [3] states:

"The justification for long term operation shall utilize the results of periodic safety review and shall be submitted to the regulatory body, as required, for approval on the basis of an analysis of the ageing management programme, to ensure the safety of the plant throughout its extended operating lifetime."

An AMP covers the systematic arrangements that are aimed at providing adequate consideration of ageing management measures, specifically engineering, operations and maintenance actions to limit the ageing degradation of SSCs to acceptable levels. An AMP is a set of plant activities relating to the understanding, prevention, detection, monitoring and mitigation of a specific ageing effect on a structure, component or group of components. Effective ageing management for SSCs includes maintenance, in-service inspection, testing and surveillance activities, with the goal of improving the reliability of SSCs (see Safety Reports Series No. 82 (Rev. 1) [6]).

In general, ageing effects on SSCs can be managed by AMPs and/or existing plant programmes. Examples of programmes to manage ageing, as described in Ref. [6], include the following:

(a)   Plant specific AMPs;
(b)   Plant specific maintenance strategies and programmes, including active components;
(c)   Activities resulting from TLAA revalidation;
(d)   Activities for preserving equipment qualification;
(e)   Plant specific programmes and procedures;
(f)   IGALL based AMPs, either fleetwide or NPP specific.

Several ways to accomplish ageing management include the following AMP types, as described in Ref. [6]:

(a)   Degradation mechanism specific AMPs (e.g. flow accelerated corrosion, stress corrosion cracking, thermal ageing);
(b)   Structure or component specific AMPs (e.g. containment, reactor coolant pumps, control rod drive housing);
(c)   General AMPs (e.g. in-service inspection, chemistry).

A list of AMPs is provided in appendix I of Ref. [6]. AMPs are divided into three groups — AMPs for mechanical components, AMPs for electrical and I&C components, and AMPs for civil structures.

Ageing management activities, as described in SSG-48 [2], can be specific to an SSC (e.g. reactor pressure vessel, pressurizer, reactor coolant pump, concrete containment structure) or commodity group (e.g. motors, valves, cables, pumps). They are typically included in either the plant specific AMPs or in the following plant programmes, which are essential to ageing management and evaluation for LTO:

(a) Maintenance programmes;
(b) Equipment qualification programmes;
(c) In-service inspection programmes;
(d) Surveillance programmes;
(e) Water chemistry programmes.

The identification of programmes to manage relevant ageing effect and degradation mechanisms for in-scope SSCs represents one of the outputs of the AMR.

The programme for LTO demonstrates that ageing effects will be adequately managed for each in-scope SSC in such a way that their intended function(s) will be maintained throughout the planned period of LTO. Therefore, the existing plant programmes used for ageing management and existing AMPs are reviewed to ensure that they will remain effective in managing the effects identified for the planned period of LTO (see para. 7.26 of SSG-48 [2]).

A systematic review of the existing plant programmes ensures that all required activities related to ageing management are implemented and effective. The recommended method of review and evaluation of AMPs and plant programmes is to check their consistency with the nine attributes of an effective AMP as described in SSG-48 [2]. A detailed description of how to perform the consistency review for the above mentioned plant programmes is given in Ref. [6].

The review and update of AMPs and existing plant programmes for the planned period of LTO is performed within the ageing management or LTO programme based on the results of the AMR. This review can lead to the following:

(a) Confirmation that AMPs and existing plant programmes are effective at managing ageing effects for the planned period of LTO;
(b) Modification of existing AMP(s) and/or plant programmes necessary to ensure fulfilment of intended safety function(s) of in-scope SSCs;
(c) Need for, and development of, new AMP(s) and/or plant programmes.

Reasons for the modification of existing AMPs or the development of new AMPs resulting from AMR conclusions could include the following:

(a) The original scope of AMPs and plant programmes does not cover all in-scope SSCs for LTO;

(b) Ageing effects of in-scope SSCs for LTO are not identified and/or managed by the existing AMP;

(c) Missing, inadequate or ineffective activities related to the existing ageing effects (e.g. monitoring, prevention, mitigation) are identified;

(d) Insufficient activities related to the existing ageing effects (e.g. periodicity of inspection, condition monitoring performed on number of samples) are identified;

(e) Incomplete or missing documentation of in-scope SSCs is revealed;

(f) Insufficient knowledge of operating history is identified;

(g) Qualified life is not documented for the period of LTO;

(h) An insufficient number of samples in surveillance programmes for the period of LTO is identified;

(i) Newly discovered ageing effects or more detailed analyses of existing ones are revealed;

(j) Missing plant procedures, methodologies and/or guidelines are identified.

As a next step, corrective measures to address deficiencies identified in the review of AMPs are proposed, and an action list is developed. Proposed corrective actions can vary in nature, such as:

(a) Development of a new AMP;

(b) Development of new plant procedures, methodologies and/or guidelines;

(c) Modification and/or replacement of in-scope SSCs;

(d) Modification of the existing AMP concerning programme scope (SSC), detection methods, condition monitoring, preventive and mitigation activities;

(e) Improvement of implementation methodology and/or control of activities of the existing AMP;

(f) Modification of operation regimes;

(g) Performance of new analyses and/or tests.

Based on the proposed corrective measures and the corresponding action list, an LTO action plan is developed and implemented.

As the main goal of many plant programmes is to manage ageing, activities undertaken as part of the programme for LTO can significantly inform and/or support the review of safety factor 4. The following aspects need to be considered when coordinating PSR and LTO activities:

(a) All in-scope SSCs are evaluated and reported when reviewing relevant safety factors;
(b) The review of safety factor 4 is performed on a component level, not on a programme level;
(c) Periodic review and reassessment of AMP effectiveness is included in the LTO programme;
(d) Periodic evaluation of plant programmes and AMPs is performed to pursue ageing management throughout the period of LTO;
(e) The existing plant programmes credited for LTO fully cover all attributes of an effective AMP.

3.1.3.2. Revalidation of TLAAs

The objective of revalidation of TLAAs is to demonstrate that ageing effects for considered in-scope SSCs will be adequately managed throughout the intended period of operation (see para. 5.67 (c) of SSG-48 [2]). This partly overlaps with the objective of safety factor 4 (see para. 5.46 of SSG-25 [1]). The results of revalidations of TLAAs are documented in safety factor 4.

The design and safe operation of some in-scope SSCs may rely upon time limited assumptions with respect to ageing, which remain valid throughout the original design life of the plant. In the case of LTO, the validity of TLAAs is reassessed, extending to cover the entire period of LTO, as described in paras 5.64–5.69 of SSG-48 [2]. The evaluation of TLAAs involves two time dependent parameters, as described in para. 5.66 of SSG-48 [2]. The first parameter is the independent variable (e.g. neutron fluence, number of reactor startups, shutdowns) and is determined from the operating history of the plant. The second parameter accounts for the ageing effects (e.g. radiation embrittlement, cumulative usage factor).

The assessment of TLAAs is typically not part of the scope of a PSR based on SSG-25 [1]. However, a review of their revalidation is included if the PSR is to be used in support of LTO (see para. 3.6 of SSG-25 [1]). The structures and components subject to revalidation of TLAAs are identified during the scope setting process. A similar approach could be followed if a PSR is used to support an LTO assessment, as shown in figures 3 and 4 of SSG-48 [2]. In addition, the LTO may generate additional time limited assumptions that need to be considered for the entire period of LTO (see para. 5.65 of SSG-48 [2] and section 7.2.3 of IAEA Nuclear Energy Series No. NP-T-3.24, Handbook on Ageing Management for Nuclear Power Plants [10]). Revalidating TLAAs demonstrates that all in-scope SSCs are capable of performing their intended safety functions throughout the whole period of LTO. Since TLAAs involve several time limited assumptions, their revalidation is closely related to those safety factors of a PSR that focus on time dependent aspects, of which ageing is the most prominent.

Revalidation of TLAAs covers the intended period of LTO, which may be equal to or longer than the time span covered by a regular PSR (ten years). This means that the results obtained from the revalidation of TLAAs would also be valid for the time span covered by the PSR.

3.1.3.3. *Technological obsolescence*

Many utilities are facing increasing demands related to technological obsolescence. Although this type of obsolescence does not manifest as physical ageing, it can result in the following:

(a) Unreliable or unavailable safety related SSCs;
(b) Commercial challenges from stretched or outsourced supply chains or higher prices;
(c) A shrinkage in the capabilities of industry and suppliers, and loss of diversity or loss of quality assurance of nuclear product lines;
(d) An increased burden associated with the management of modifications.

As part of safety factor 4, para. 5.48 of SSG-25 [1] identifies the obsolescence of technology used in the NPP as one of the technical aspects to be considered in this review.

The attainment of LTO requires an in depth review of ageing management. Part of this review concerns the effectiveness of the proactive identification of obsolescence in advance and the corrective actions taken to address it. A comprehensive technological obsolescence management programme is expected to be in place to identify, prioritize and implement solutions to the obsolescence of SSCs, particularly those important to safety.

The review of the following safety factors needs to consider technological obsolescence:

(a) Safety factor 2 (actual condition of SSCs important to safety): Paragraphs 5.27 and 5.29 of SSG-25 [1] identify the importance of knowing the current and anticipated state of obsolescence of in-scope SSCs and later list the review of this topic as a task within the scope of safety factor 2.
(b) Safety factor 4 (ageing): Paragraph 5.48 of SSG-25 [1] includes an evaluation of the impact from the obsolescence of technology. An obsolescence management programme satisfying the guidance given in section 6 of

SSG-48 [2] will meet the intent of para. 5.48 of SSG-25 for PSR. In addition, the prioritization of obsolescence issues requires an understanding of the in-scope SSCs' reliability and failure history. Data gathered to support this review are important inputs when considering the effectiveness of AMPs.

(c) Safety factor 8 (safety performance): Paragraphs 5.86 and 5.94 of SSG-25 [1] determine whether the plant safety performance indicators and use of operating experience are effective. This review will include identification of the drivers, which may include obsolescence, for replacement of in-scope SSCs and the adequacy of the methodologies in place to trend, analyse and act upon these data.

(d) Safety factor 10 (organization, the management system and safety culture): Paragraph 5.114 of SSG-25 [1] recommends that the operating organization have a management system to ensure that suitably qualified human resources are available and that this system includes succession planning. The obsolescence management programme can highlight instances of reduced or absent technical capability from suppliers or the wider industry for which an effective mitigation might be the training and retention of knowledge within the utility or supporting organizations.

3.1.4. Organization, the management system and safety culture (safety factor 10)

3.1.4.1. Management system for LTO

A management system is essential for ensuring the protection of people and the environment from the harmful effects of ionizing radiation. An effective management system integrates all elements of management so that requirements for safety are established and applied coherently with other requirements, including those necessary for efficient and safe LTO.

An effective AMP applied at all stages of the lifetime of an NPP and carried out within an effective organizational structure that includes sufficient resources and an adequate management system supported by a strong safety culture are fundamental prerequisites for safe LTO.

The operating organization responsible for PSR is required to develop, implement, assess and continuously improve a management system, which includes quality management, in accordance with the requirements established in IAEA Safety Standards Series No. GSR Part 2, Leadership and Management for Safety [11].

The operating organization is expected to adopt a comprehensive organizational arrangement to prepare and implement the programme for LTO, in addition to existing processes associated with ageing management consistent with the regulatory requirements relevant to LTO. These individual aspects of ageing management related to organization, management and safety culture are assessed within the PSR. The results of the review of these aspects provide the supporting evidence for the justification of LTO: that the plant can be operated safely beyond the original time frame established in the licence conditions, design limits, safety standards and/or regulations.

The following general plant management aspects are typically assessed in the PSR performed in accordance with SSG-25 [1]:

(a)   Safety policy;
(b)   Management system;
(c)   Documentation;
(d)   Human resources.

Nevertheless, some requirements and recommendations specific to non-technical aspects of LTO are not included in the standard scope of PSR. The expansion of the PSR scope is considered to be a suitable way to provide adequate supporting evidence by the following means:

(a)   Safety policy

Senior management establishes the safety policy and subsequent goals, strategies, plans and objectives for the organization. Adherence of the safety policy to IAEA Safety Standards is typically a review area of safety factor 10 in the PSR performed according to para. 5.113 of SSG-25 [1]. The assessment of safety factor 4 of the PSR provides additional assurance that a systematic, effective and comprehensive AMP and subsequent procedures for ageing management are in place.

A policy for LTO or a high level programme or strategy fulfilling the policy functions is expected to be in place and consistent with the overall safety policy. Specific guidance in the PSR basis document for reviewing the policy (and associated programme or strategy) for LTO is suggested.

(b)   Management system

The management system forms the integrated set of interrelated or interacting components for establishing policies and objectives and enabling the organization's objectives to be achieved in an efficient manner. The following key components are specified in the management system:

— Organizational structure;
— Processes;
— Responsibilities and accountabilities;
— Levels of authority;
— Interfaces within an organization and with external organizations.

The dedicated organizational structures to perform evaluations for LTO established by the operating organization are covered by the existing integrated management system. The comprehensive review of the management system's adequacy is the objective of the safety factor 10 review in the PSR, as stated in para. 5.112 of SSG-25 [1].

For PSR supporting LTO, a focus on specific LTO aspects, such as the definition of additional responsibilities, authorities and human resources required, is necessary.

Establishing effective communication channels and good working relations with the regulatory authority, and ensuring common understanding of, and compliance with, safety requirements and their interface with other requirements, is necessary for the LTO process.

As preparing a plant for safe LTO usually includes several important safety modifications and refurbishments, the plant will need to follow a configuration management programme or modification management programme that reflects its evolving status. Operating organizations usually establish an entity to maintain the integrity of the plant's design. In-depth review of the configuration management function and modification management programme within PSR provides important reassurance that the design integrity will be correctly maintained in the LTO process, and that the modification management process will ensure proper implementation of important modifications resulting from the ageing management and LTO assessment. Configuration management and modification management review is included in the standard scope of PSR (see para. 5.114 of SSG-25 [1]) and remains valid for LTO.

(c) Documentation

The organizational entity responsible for maintaining design integrity is also responsible for ensuring that the knowledge of information related to the design basis is also maintained. The PSR should focus on the availability of complete and up to date design basis documentation to the operating organization to support appropriate configuration management and modification management and to allow identification of the TLAAs for the plant. If the PSR identifies that the design basis documentation is insufficient or obsolete, appropriate corrective actions, including reconstitution of the design basis, might need to be taken. The design basis information is typically embodied in the updated FSAR and the supporting design basis documentation.

A data collection and record keeping system is necessary from the early stages of the lifetime of the plant to support ageing management. Design documentation, including documentation from suppliers, is essential in supporting effective ageing management. The adequacy of the control of documents and records, as well as their retrievability, is reviewed in the PSR (see para. 5.114 of SSG-25 [1]), with special focus on those records essential for ageing management and LTO.

(d) Human resources

The operating organization is responsible for ensuring access to all necessary competencies (in house or external) and resources to safely conduct activities. Expertise in the technical, human and organizational aspects of the facility and safety is usually sustained in house. LTO imposes an additional burden on personnel responsible for maintaining design integrity, design safety assessment and ageing management.

Ensuring the competence of plant personnel, who are familiar with LTO, its principles and concepts and understand the effects of ageing on in-scope SSCs and the obsolescence of equipment, is important for safety. Adequate training is expected to be provided for the operations, maintenance and engineering staff in these areas.

Another important aspect to consider for LTO is the loss of experienced staff as they reach retirement age, which typically occurs after 30–40 years of service. Effective knowledge management is therefore an essential prerequisite for prolonged plant operation.

Sufficient human resources, adequate training and knowledge management are essential for implementation of LTO. LTO may require additional aspects (e.g. identification of additional human resource requirements, associated training requirements and required pipelining of human resources) to be assessed as part of the review of safety factor 10 (see para. 5.114 of SSG-25 [1]). Policies and processes for maintaining know-how are reviewed in safety factor 12 (see para. 5.129 of SSG-25 [1]). Nevertheless, an enlarged scope and focus of the review will be needed to address issues specific to LTO, and in this respect the assessment results performed within the LTO programme will be important.

3.1.5. Summary of key inputs from LTO to PSR

If the PSR is used to extend a plant's operating licence, the evaluation of safety factors 2 to 5 needs to consider the entire planned period of LTO (see para. 7.38 of SSG-48 [2]). Therefore, inputs from the LTO assessment to the PSR are very important. As the LTO assessment and the PSR programme often do not run in parallel, the following inputs from LTO to PSR can be considered:

(a) Corrective actions from LTO assessments provide inputs to the PSR global assessment;

(b) The LTO action plan is fully incorporated into the subsequent PSR integrated implementation plan.

The detailed inputs from the LTO assessment (topical area: review of AMPs and plant programmes) to support the review of individual safety factors in the PSR can be as follows:

(a) List of in-scope SSCs for review of relevant AMP and plant programmes (applicable to safety factors 2 to 4);

(b) List of AMPs and other programmes used to manage ageing (applicable to safety factors 2 to 4);

(c) Methods for monitoring and mitigating ageing effects (applicable to safety factor 4);

(d) Summary of the scope and results of inspections that prove the functional capability of in-scope SSCs (applicable to safety factors 2 and 4);

(e) Summary of understanding of ageing effects, including safety margins (applicable to safety factor 4);

(f) List of preventive actions for minimizing and controlling ageing effects (applicable to safety factor 4);

(g) Summary of condition monitoring, condition indicators and corresponding acceptance criteria applied to in-scope SSCs (applicable to safety factor 4);

(h) Use of operating experience and R&D results in relevant AMPs and plant programmes (applicable to safety factors 4 and 9);

(i)   Management of records and documentation in relevant AMPs and plant programmes (applicable to safety factors 4 and 10);

(j)   Management of plant procedures that have interconnections with relevant AMPs and plant programmes (applicable to safety factors 4 and 11).

The scope of the review of the safety factors needs to be adapted to determine the feasibility of LTO.

3.2. KEY OUTPUTS FROM PERIODIC SAFETY REVIEW FOR JUSTIFICATION OF LONG TERM OPERATION

3.2.1. Obsolescence assessment

As discussed in Section 3.1.3, obsolescence is the non-physical ageing of SSCs that are becoming out of date in comparison with current knowledge, codes, standards and regulations, and technology.

Conceptual aspects of obsolescence, such as obsolescence of knowledge and compliance with current regulations, codes and standards, are addressed in Requirements 5 and 12 of SSR-2/2 (Rev. 1) [3], which deal with safety policy and PSR, and in safety factors 2 and 8 according to SSG-25 [1], which deal with actual condition of in-scope SSCs and safety performance.

SSG-25 [1] provides a framework for the review of obsolescence management as part of an AMP for LTO, as follows:

(a)   Paragraphs 5.29 and 5.30, concerning the examination of ageing and obsolescence, as part of the review of safety factor 2;

(b)   Paragraph 5.48, concerning obsolescence management as part of ageing management for LTO, included in the review of safety factor 4;

(c)   Paragraph 5.86, concerning the evaluation of the occurrence of obsolescence issues and the impact on safety performance, included in the review of safety factor 8.

Furthermore, para. 6.1 of SSG-48 [2] states: "Technological obsolescence of the SSCs in the plant should be managed through a dedicated plant programme with foresight and anticipation and should be resolved before any associated decrease in reliability and availability occur[s]."

In addition, para. 7.3 of SSG-48 [2] states: "The organizational arrangements for the management of physical ageing, including technological obsolescence, should be properly implemented and should be one of the prerequisites for a decision to pursue long term operation of the nuclear power plant." The decision of an operating organization to pursue LTO is typically based on an evaluation (feasibility study) that includes an evaluation of past operating experience at the plant relating to ageing, obsolescence and other safety issues.

TABLE 4. TYPES OF OBSOLESCENCE *(reproduced from SSG-48 [2])*

Type of obsolescence	Description	Consequences	Management
Technological obsolescence	• Lack of spare parts and technical support • Lack of suppliers • Lack of industrial capabilities	• Declining plant performance and safety due to increasing failure rates and decreasing reliability	• Systematic identification of useful service life and anticipated obsolescence of SSCs • Provision of spare parts for planned service life and timely replacement of parts • Long term agreements with suppliers • Development of equivalent structures or components
Regulations, codes and standards obsolescence	• Deviations from current regulations, codes and standards for structures, components and software • Design weaknesses (e.g. in equipment qualification, separation, diversity or capabilities for severe accident management)	Plant safety level below current regulations, codes and standards (e.g. weaknesses in defence in depth or higher risk of core damage)	Systematic reassessment of plant safety against current regulations, codes and standards (e.g. through PSR) and appropriate upgrading, backfitting or modernization
Knowledge obsolescence	Knowledge of current regulations, codes and standards and technology relevant to SSCs not kept current	Opportunities to enhance plant safety missed	Continuous updating of knowledge and improvement of its application

In accordance with table 1 in SSG-48 [2], obsolescence has been divided into three broad categories, namely:

(a)   Technology;
(b)   Regulations, codes and standards;
(c)   Knowledge.

Where the PSR is to be used in decision making for LTO, the evaluation of obsolescence needs to be an area of specific focus to verify whether NPP safety can be impaired if the obsolescence of SSCs is not identified and rectified in a timely manner.

PSR can be a very useful tool in supporting the justification of LTO, especially in the evaluation of obsolescence of both codes and standards and knowledge.

NPP safety can be impaired if the obsolescence of SSCs is not identified in advance and corrective actions are not taken before the associated decrease in the reliability or availability of SSCs occurs. The management of obsolescence forms part of the general approach for enhancing NPP safety through improvements in both the performance of SSCs and safety management.

These obsolescence types, including a description, consequences and the management approaches to prevent or mitigate them, are shown in Table 4.

Aspects of obsolescence are addressed in various safety factors in SSG-25 [1], as follows:

(a)   Safety factor 2 (actual condition of SSCs important to safety)

It is essential to document the condition of each in-scope SSC thoroughly. Additionally, knowledge of any existing or anticipated obsolescence of plant systems and equipment needs to be considered as part of this safety factor.

The review of the actual condition of the in-scope SSCs also includes the examination of the following aspects for each SSC:

— The current state of the SSC with regard to its obsolescence;
— The dependence on obsolete equipment for which no direct substitute is available.

The actual condition of the in-scope SSCs is reviewed considering ageing processes, obsolescence of plant systems and equipment, modification history and operating history.

(b) Safety factor 8 (safety performance)

The review of safety performance evaluates whether

— The plant has appropriate processes in place for the routine recording and evaluation of safety related operating experience, including replacements of in-scope SSCs owing to failure or technological obsolescence;
— The adequacy of the plant's safety performance methodologies and processes, including trend analyses regarding component replacements owing to failure or obsolescence.

Subtopics of obsolescence of knowledge include the following:

— Knowledge management: Building, collecting, transferring, sharing, preserving, maintaining and utilizing knowledge are essential to developing and keeping the necessary technical expertise and competences required for nuclear power programmes and other nuclear technology.
— Competence management: Competence is the combination of knowledge, skills and attitudes needed by a person to perform a particular job. Management commitment ensures appropriate competence of the workforce (professional, competent, versatile and motivated).
— Competence policy: This covers vision, goals and strategies, and plans for their delivery.
— Planning: This involves review of functions, systematic assessment of competence and determination of the size and composition of staff, training and development, as well as use of external support to fill competence needs.
— Responsibilities: The definition of responsibilities includes assigning a first line manager for staff competence development, designating the number of personnel for competence management and the training coordinator responsible for identifying competence gaps, and assigning responsibility for recruitment, training or outsourcing.
— Documentation: Competences for each task (job specifications), competences possessed by individuals and their competence development plans, staff qualification and training records are adequately documented.
— Knowledge retention: This includes the structuralized process of interviewing and documenting the experience of more experienced specialists; strategic planning before the retirement of employees; and

allowing less experienced employees to work under the supervision of more experienced ones.

— Configuration management: This is the process of identifying and documenting the characteristics of the SSCs of a facility and of ensuring that changes to these characteristics are properly developed, assessed, approved, issued, implemented, verified, recorded and incorporated into the facility documentation.

— Succession planning: Succession planning is a systematic approach by an organization to save leadership permanence in crucial positions, by means of intellect retention and development, retaining knowledge capital for the future and encouraging individual progression. Identifying and developing employees to ensure that key positions can be filled with qualified internal candidates (or, when necessary, that candidates can be recruited externally) in advance of actual needs is important.

A mapping of obsolescence to various safety factors is shown in Table 5. Practically all safety factors are affected, although levels of impact differ.

TABLE 5. MAPPING OF OBSOLESCENCE TO VARIOUS SAFETY FACTORS

Safety factor		Extent	Area, scope
SF1	Plant design	****[a]	Design weaknesses (e.g. defence in depth, independence, diversity, provisions for design extension conditions) Deficiencies of the plant design against latest codes and standards
SF2	Actual condition of SSCs important to safety	**	Technology related obsolescence Current in-scope SSCs' status relating to technological obsolescence Dependence on obsolete equipment due to loss of supply chain
SF3	Equipment qualification	*	Design weaknesses in equipment qualification, loss of supply chain

TABLE 5. MAPPING OF OBSOLESCENCE TO VARIOUS SAFETY FACTORS (cont.)

Safety factor		Extent	Area, scope
SF4	Ageing	**	Technology related obsolescence Systematic identification of anticipated technological obsolescence of SSCs
SF5	Deterministic safety analysis	*	Deficiencies in provisions to cope with design extension conditions Deficiencies in severe accident management
SF6	Probabilistic safety assessment	*	Deficiencies resulting from the review of hazard analysis (site characteristics, external hazards, internal hazards, combination of hazards[b])
SF7	Hazard analysis	**	
SF8	Safety performance	*	Adequacy of the plant safety performance methodology used to assess obsolescence
SF9	Use of experience from other plants and research findings	*	Deficiencies in evaluation of past operating experience at other plants relating to ageing, obsolescence and other safety issues
SF10	Organization, the management system and safety culture	***	Deficiencies due to obsolete standards and regulations Procedures and documentation of the management systems (e.g. configuration, knowledge obsolescence) Deficiencies in commitment to safety
SF11	Procedures	**	Deficiencies in effectiveness of procedures compounding potential knowledge obsolescence deficiencies
SF12	Human factors	***	Deficiencies in understanding obsolescence management, staffing, competence, knowledge retention, management and training
SF13	Emergency planning	*	Deficiencies in capabilities for severe accident management in emergency planning

TABLE 5. MAPPING OF OBSOLESCENCE TO VARIOUS SAFETY FACTORS (cont.)

Safety factor		Extent	Area, scope
SF14	Radiological impact on the environment	*	Deficiencies in emergency planning and radioactive discharge management to minimize risk to workers, the public and the environment

[a] *: low impact; **: medium impact; ***: high impact.
[b] Codes and standards may change as a consequence of new identified hazards or revision of existing hazards.

Technological obsolescence is usually covered by the LTO programme. The plant programme to manage technological obsolescence is considered a precondition for LTO (see section 6 of SSG-48 [2]) and needs to be consistent with the nine generic attributes of an effective AMP (see table 2 of SSG-48 [2]). For conceptual aspects of obsolescence, SSG-48 [2] refers to the PSR. According to Table 5, the main contribution from the PSR to support the justification of LTO regarding both codes and standards obsolescence would be provided primarily from safety factor 1 and safety factor 7. A review of obsolescence of knowledge would primarily be provided from safety factor 10 and safety factor 12. Other safety factors may also contribute to the review of obsolescence.

3.2.2. Cumulative effects of physical ageing and non-physical ageing

Evaluation of the consequences of the cumulative effects of both ageing and obsolescence on the safety of an NPP is a continuous process and is required to be assessed in a PSR or an equivalent safety assessment under alternative arrangements (see paras 4.6–4.8 of SSG-48 [2]).

Two different aspects of ageing need to be assessed: both physical and non-physical (obsolescence). Physical ageing is a process by which the physical properties of SSCs gradually deteriorate with time. Non-physical ageing is where SSCs are no longer up to date (obsolete) as a result of the obsolescence of technology (e.g. termination of production, unavailability of suppliers, end of spare parts production). One of the prerequisites for a decision to pursue LTO of an NPP, as required in para. 7.3 of SSG-48 [2], is that the organizational arrangements for the management of physical ageing and obsolescence should be properly implemented.

Although safety factor 1 and safety factor 3 do not explicitly mention this, the ageing in combination with obsolescence may result in plant modifications as corrective actions in the context of safety factor 1 (plant design). Cumulative aspects of physical ageing and obsolescence are addressed in the following safety factors:

(a) Safety factor 2 (actual condition of SSCs important to safety)

The review of safety factor 2 is expected to contain relevant information for managing the cumulative aspects of ageing and obsolescence. Evaluation of the cumulative aspects of ageing and obsolescence can be significantly supported by systematic monitoring and reporting of the health of systems and components, which is based on the performance and conditions of in-scope SSCs, in combination with a proactive obsolescence management programme.

(b) Safety factor 3 (equipment qualification)

The evaluation of safety factor 3 may also identify obsolescence (e.g. equipment qualification is no longer valid because of a lack of qualified spare parts), which may have a cumulative impact on the general conditions of SSCs. Therefore, an integrated evaluation approach is recommended to identify potential obsolescence issues in equipment qualification and respective corrective measures for LTO.

(c) Safety factor 4 (ageing)

The ageing management methodology applied for the evaluation of safety factor 4 follows the recommendations on management of technological obsolescence, as provided in section 6 of SSG-48 [2]. The scope of the AMP is expanded to include assessment of physical and non-physical ageing phenomena. The evaluation may result in identification and implementation of corrective measures for LTO.

(d) Safety factor 8 (safety performance)

Complementary to the ageing management for LTO included in the review of safety factor 2 and safety factor 4, the evaluation of safety factor 8 may identify negative trends in the replacement of in-scope SSCs owing to obsolescence. The implementation of corrective measures, as well as an enhancement of the obsolescence management programme, may be required for LTO.

(e) Safety factor 10 (organization, the management system and safety culture)

A strong safety culture is considered an integral part of an NPP's management system, which conforms to the attributes provided in IAEA Safety Standards Series No. GS-G-3.1, Application of the Management System for Facilities and Activities [12], and listed below:

— Safety is a clearly recognized value;
— Leadership for safety is clear;
— Accountability for safety is clear;
— Safety is integrated into all activities;
— Safety is learning driven.

An inadequate safety culture may lead to:

— Insufficient focus on safety aspects;
— Lack of external operating experience use in the company;
— Poor performance of internal experience feedback, showing lack of due event analysis and an inefficient programme for corrective actions;
— Degraded knowledge management with insufficient motivation mechanisms for knowledge sharing and transfer from senior experts.

The review of safety culture as a part of PSR safety factor 10 for LTO could provide a comprehensive overview of all key aspects and attributes of a strong safety culture (see para. 2.36 of GS-G-3.1 [12]).

3.2.3. Identification of safety improvements

In addition to the continuous safety verification process, PSR provides, through a comprehensive review, a sound basis to determine whether the plant is operated in conformance with the current licensing basis and whether the plant safety levels can be reasonably enhanced by means of the implementation of further safety improvements.

PSRs are typically conducted at ten year intervals until the end of operation. During that time period, developments in safety standards and operating practices may be made that go beyond the current licensing basis. In that case, the PSR will need to demonstrate the extent to which the plant conforms to current national and/or international safety standards and operating practices for existing plants.

Additionally, whereas the continuous safety verification process is typically focused on specific and the most recent plant experience, the PSR provides a comprehensive evaluation of the long term cumulative effects of plant operating experience and SSC performance. In this sense, the PSR needs to demonstrate that the arrangements that are in place for both safety activities and the care of the SSCs installed at the plant are effective and adequate to ensure plant safety until at least the next PSR and, if appropriate, until the end of planned operation.

3.2.3.1. *Inputs to identify PSR safety improvements*

Paragraph 4.44 of SSR-2/2 (Rev. 1) [3] states:

"Safety reviews shall address, in an appropriate manner: the consequences of the cumulative effects of plant ageing and plant modification; equipment requalification; operating experience, including national and international operating experience; current national and international standards; technical developments; organizational and management issues; and site related aspects."

In identifying safety improvements, all findings from the safety factor reviews are expected to be considered. The negative findings or deviations identified in the PSR will lead to safety improvements. When using the PSR in support of justification of LTO, the conclusions and findings from the safety factors covered in Section 3.1 need to be evaluated in greater detail, as they cover those areas related to managing the effects associated with LTO.

Although not directly related to LTO, PSR findings from other safety factors may also lead to significant safety improvements that will support justification of LTO. One of the most significant activities within a PSR is the comprehensive analysis of national and international good practices that go beyond the existing plant licensing basis, as these assessments represent a primary source of identification of negative PSR findings or deviations and will support plants to further improve safety levels.

It is essential that national and international good practices are identified in the early stages of the PSR process, and these need to be evaluated comprehensively during the safety factor review stage. Non-regulatory documents from national and international organizations such as the IAEA, regulatory bodies, owners' groups, nuclear operator associations and technical institutes are relevant sources to be used in evaluating the plant against national and international good practices.

3.2.3.2. *Safety improvements identified in the PSR for LTO*

The safety improvements identified in the PSR are expected to be submitted for evaluation by the regulatory body within the regulatory process prior to LTO and implemented in accordance with a schedule agreed to with the regulatory body. Those safety improvements aimed at ensuring that ageing effects for the planned period of LTO are effectively managed need to be scheduled to ensure measures are implemented before the period of LTO. Implementation of these safety improvements could also be planned during the period of LTO. However, this needs to be justified by the operating organization and, if required, made subject to approval by the regulatory body.

The safety improvements directly related to ageing effects arising from LTO generally rely on:

(a) Development and implementation of plant programmes to effectively detect, monitor, assess and correct ageing effects associated with LTO;
(b) Performance of inspection, testing and monitoring activities on in-scope SSCs to detect ageing effects;
(c) Performance of actions to mitigate detected ageing effects, including improvements in operations and maintenance;
(d) Repair and replacement of in-scope SSCs;
(e) Performance of activities required to revalidate TLAA assumptions.

PSR in support of LTO will not only justify that current safety levels are adequate for LTO at the time the PSR is performed, but also show that adequate arrangements are already in place to effectively manage the future effects of LTO until the end of its intended period.

Safety improvements in support of LTO are not limited to those aspects directly related to the effects associated with LTO (areas covered in Section 3.1). Scheduling of these other improvement actions would be based on their safety significance with possible additional considerations, including the amount of allocated resources or the potential cross-cutting impact of the improvements.

Identification of non-conformances with a licensing basis or design basis as a result of a PSR is not considered a safety improvement and their resolution is generally managed through the regulatory processes established to that effect within the current licensing basis.

3.2.3.3. *PSR outputs necessary in support of justification of LTO*

Paragraph 1.3 of SSR-2/1 (Rev. 1) [9] states:

"it is expected that a comparison will be made with the current standards, for example as part of the periodic safety review for the plant, to determine whether the safe operation of the plant could be further enhanced by means of reasonably practicable safety improvements."

To address whether the resolution of some of the negative findings or deviations is unreasonable and not practicable, an evaluation of the risks associated with not addressing the findings needs to be performed. Although risk evaluations of the findings from a PSR for continued operation could benefit from the use of a PSA, these evaluations do not generally rely solely on probabilistic analysis, and different approaches may be used, including DSA, engineering judgement or cost–benefit analysis (see para. 6.7 of SSG-25 [1]).

When performing a risk evaluation for continued operation, special consideration will be given to the time necessary for implementing the safety improvement(s) and the remaining planned lifetime of the plant. The actual duration of the benefit obtained from the safety improvement(s) needs to be considered, as well as whether the benefit of the safety improvement(s) will remain during the transition of the plant to the decommissioning stage.

An integrated implementation plan is developed as part of the PSR, considering all the reasonably practicable safety improvements. An integrated implementation plan typically includes the following information for each safety improvement:

(a) Identification and name of potential improvement;
(b) Reference to related PSR finding(s);
(c) Detailed scope of the activity;
(d) Priority of the activity as evaluated during the global assessment process;
(e) Implementation deadline;
(f) Coordinator;
(g) Resources required for implementation.

To more efficiently manage the integrated implementation plan, safety improvements may also be grouped in accordance with the areas covered, resources used or type of activity.

3.2.4. Environmental impact of LTO

In some Member States, an environmental impact assessment is required before entering into LTO. This section provides an overview of where outputs from a PSR could be used to support LTO regarding an environmental impact assessment.

This section is intended to identify the information related to the different safety factors of a PSR to assess the environmental impact of LTO, as required by national regulations[5]. The environmental impact can be associated with the following safety factors:

— Safety factor 1: Plant design;
— Safety factor 8: Safety performance;
— Safety factor 14: Radiological impact on the environment.

(a)   Safety factor 1: Plant design

The scope of this safety factor is defined in para. 5.17 of SSG-25 [1]; within that paragraph, the following aspects related to environmental impact are identified.

— Plant modifications: The operating organization may identify potential new sources of radiological impact through examining plant changes and the actual condition of SSCs.
— Strategy for the spent fuel storage: When conducting the PSR in support of justification of LTO, this review needs to consider the anticipated generation of spent fuel during the expected period of LTO and the capacity of current storage facilities to safely manage it, together with contingency plans for the different assumptions considered. Evaluation of the strategy for spent fuel storage also considers the effect of the condition of the spent fuel and associated storage facilities, based on engineering assessments and the inspection and monitoring activities performed during the period of PSR.

(b)   Safety factor 8: Safety performance

The scope of this safety factor is defined in para. 5.86 of SSG-25 [1]; within that paragraph, the following aspects related to the environmental impact are identified:

— Off-site contamination and radiation levels;
— Discharges of radioactive effluents;
— Generation of radioactive waste.

This safety factor examines specific data on radiation doses and radioactive effluents and the effectiveness of the radiation protection measures in place

---

[5] The scope and criteria for environmental impact assessments are typically contained in national regulations.

to ensure that radioactive effluents are properly managed and kept within prescribed limits.

Paragraph 5.91 of SSG-25 [1] states:

> "Data on the generation of radioactive waste should be reviewed to determine whether operation of the plant is being optimized to minimize the quantities of waste being generated and accumulated, taking into account the national policy on radioactive discharges and international treaties, standards and criteria, etc."

All those aspects are analysed in preparation for LTO.

(c)   Safety factor 14: Radiological impact on the environment

Paragraph 5.146 of SSG-25 [1] states: "The operating organization should have in place an established and effective monitoring programme that provides data on the radiological impact of the nuclear power plant on its surroundings." This monitoring programme needs to ensure that emissions and discharges are adequately controlled and are as low as reasonably possible. This has to be taken into account both in the preparation for LTO and during LTO. Data collected in the radiation monitoring programme include data on radionuclide concentrations in air, water (including river water, sea water and groundwater), soil, agricultural and marine products, and wild flora and fauna. The operating organization may identify potentially new sources of radiological impact by examining relevant plant modifications and the actual condition of in-scope SSCs.

The monitoring programme is reviewed for LTO to confirm whether it remains appropriate and sufficiently comprehensive to demonstrate that the radiological impact of the plant on the environment remains within the prescribed limits for the period of LTO. In some Member States, the environmental impact assessment of a major project, such as LTO of a nuclear facility, is subject to different regulations and follows a process that is independent and separate from PSR.

4. GLOBAL ASSESSMENT

This section outlines a method to both consolidate the results and evaluate the findings from the review of individual safety factors, as well as additional activities included in the PSR scope for LTO.

4.1. PURPOSE OF GLOBAL ASSESSMENT

The global assessment considers all findings and proposed safety improvements based on the safety factor reviews and interfaces among different safety factors (e.g. para. 2.17 of SSG-25 [1]).

In accordance with SSG-25 [1], the global assessment:

(a) Is performed by an interdisciplinary team with appropriate expertise in operation, design and safety at the plant, including some participants from the safety factor reviews and some members independent from the safety factor review teams (para. 6.5 of SSG-25 [1]);
(b) Considers strengths, opportunities for improvement and associated reasonable and practicable safety improvements from the safety factor reviews (para. 6.6 of SSG-25 [1]);
(c) Applies a method for prioritizing safety improvements based on safety significance resulting from a combination of DSA, PSA, engineering judgement, cost–benefit analysis and/or risk analysis, taking into account the remaining planned lifetime of the plant;
(d) Contains an evaluation of risk associated with combined deviations, strengths and opportunities for improvement in the short term, prior to implementation and in the long term (paras 6.8, 6.9 and 6.10 of SSG-25 [1]);
(e) Reviews the adequacy of the plant's defence in depth concept (para. 6.11 of SSG-25 [1]);
(f) Is documented in the report on PSR for LTO;
(g) Presents the safety improvements that form the basis of the commitments from the plant operator to the regulatory body (paras 6.7, 6.10 and 6.12 of SSG-25 [1]).

The global assessment also provides information on the following:

(a) The grading of deviations;
(b) The analysis of interfaces, gaps and overlaps among the different safety factor reviews;

(c)   The analysis of the combined effects of deviations;
(d)   The assessment of defence in depth;
(e)   The determination of reasonably practicable safety improvements;
(f)   The development of an integrated implementation plan;
(g)   The evaluation of the current overall level of plant safety until the next PSR and, if applicable, for the proposed duration of LTO.

In cases where the PSR is used for decision making on LTO, the experts and units responsible for the plant's preparation for LTO are involved in the appropriate phases and activities performed within the global assessment.

4.2. DEVIATIONS AND THEIR SAFETY SIGNIFICANCE

As noted in paras 5.12 and 8.14 of SSG-25 [1], the safety significance of negative findings needs to be determined using deterministic and probabilistic methods in order to categorize and prioritize proposed corrective actions and improvements. A method for determining the safety significance is established prior to performing the global assessment (para. 6.7 of SSG-25 [1]) and is documented in the PSR basis document (para. 4.6 of SSG-25 [1]).

The relevance of each identified improvement could be determined based on the safety relevance of the identified deviation and the effectiveness of the proposed safety improvement in resolving the deviation. Another possibility is to use an approach based on DSA, PSA, engineering judgement, cost–benefit analysis and/or risk analysis, or a combination thereof (para. 6.10 of SSG-25 [1]) to assess the safety significance of safety improvements.

A variety of approaches could be adopted to establish the safety significance of deviations. A sound approach for judging the safety significance of findings could be based, for example, on the methodology described in Safety Reports Series No. 12, Evaluation of the Safety of Operating Nuclear Power Plants Built to Earlier Standards — A Common Basis for Judgement [13]. In addition, this Safety Report provides useful information on the prioritization of corrective measures and safety improvements.

The prioritization of corrective actions for LTO, including replacement of obsolete equipment, could use the same or a similar approach or procedure as used for the safety significance determination of PSR deviations. If the PSR is used in support of LTO, the findings from PSR and LTO could be handled as a common set of inputs into the global assessment and their safety significance determined at an early stage of the global assessment.

4.3. ANALYSIS OF INTERFACES, GAPS AND OVERLAPS

According to para. 6.3 of SSG-25 [1], an analysis of the interfaces among the various safety factors is carried out as a part of the global assessment. The approach taken uses IAEA Safety Standards Series No. SF-1, Fundamental Safety Principles [14] as a basis. The global assessment also considers overlaps and identifies omissions among the individual safety factors. It assesses whether additional commonalities among deviations are identified and whether the addition and/or optimization of the identified safety improvements is required, as well as whether the safety improvements are reasonable and practicable when taking into account the insights from all the safety factor reviews (see para. 6.6 of SSG-25 [1]).

Potential interfaces among the different safety factors are shown in appendix I, table 1 of SSG-25 [1]. The goal of the analysis of these interfaces and of individual negative findings in the global assessment is to verify that the safety factor review has used appropriate input data, considered the deviations identified in the other safety factor reviews and provided a coherent set of results and conclusions. The impact of identified deviations on input data to safety factor evaluation and implemented interfaces could be checked at this stage of global assessment. The interrelations identified among individual negative findings provide important inputs into the development of the final integrated implementation plan of corrective measures and safety improvements.

To ensure adequate management of interfaces in the assessment phase of the PSR, all findings that are related to other safety factors are distributed to the reviewers of the other safety factors within an appropriate time frame. Since the analysis of interfaces within the global assessment could be significantly time consuming, it is recommended that the global assessment team monitor the interfaces in the course of the assessment phase of the PSR.

An important input for justifying continued operation may be the analysis of high level categories consistent with SF-1 [14]. This analysis could be accomplished by synthesizing the safety factor review and the results of justification of LTO, aiming at ensuring that these safety principles are fulfilled until the next PSR cycle or the end of the expected period of LTO. Use of a national high level requirement for nuclear safety could be the most convenient way to perform this analysis.

The analysis of gaps and overlaps is to identify any missing and overlapping aspects of safety factor reviews, thereby contributing to assurance that PSR results are comprehensive and consistent. When coordinating PSR and justification of LTO, questions of completeness and consistency of results could be considered even more important.

4.4. ANALYSIS OF THE COMBINED EFFECTS OF FINDINGS

The level of plant safety is determined by a global assessment reflecting, among other things, the combined effects of all safety factors. Paragraph 4.22 of SSG-25 [1] states that "[i]t is possible that a negative finding (deviation) in one safety factor can be compensated for by a positive finding (strength) in another safety factor." Although negative findings may be individually acceptable, their combined effects are reviewed and verified for their acceptability (para. 6.9 of SSG-25 [1]).

The identification of the safety significance of each individual deviation does not reflect the possible influence of positive findings (strengths) or other compounding deviations resulting from other safety factors. The cumulative effect of synergic effects of positive and negative findings on other deviations is assessed in the global assessment.

The cumulative effect of findings from LTO and PSR could also be considered in this process, since synergies of findings from these sources could exist. Failure to include all relevant findings from different sources could lead to an underestimation of significant impacts on plant safety.

The global issues representing the possible result of the combined effects of findings could be identified in the global assessment and considered in the preparation of the integrated implementation plan. These global issues could indicate a system deficiency, the occurrence of which has not been identified from the safety factor reviews but emerges from a set of interrelated deviations.

A multidisciplinary expert team, carrying out the analysis of combined effects, may include members with expertise in the following areas:

(a) Safety assessment (DSA, PSA);
(b) Actual plant conditions;
(c) Ageing management;
(d) Radiation protection;
(e) Design (design basis and design requirements);
(f) Accident management;
(g) Operation;
(h) Human factors;
(i) Organization and management.

4.5. ASSESSMENT OF DEFENCE IN DEPTH

The global assessment reviews the extent to which the safety requirements relating to the concept of defence in depth and the fundamental safety functions (i.e. reactivity control; removal of heat from the reactor core and the fuel store; and confinement of radioactive material, shielding against radiation and control of planned radioactive releases) are fulfilled (para. 6.11 of SSG-25 [1]).

Requirement 13 of IAEA Safety Standards Series No. GSR Part 4 (Rev. 1), Safety Assessment for Facilities and Activities [15] states that **"It shall be determined in the assessment of defence in depth whether adequate provisions have been made at each of the levels of defence in depth."**

Paragraph 4.46 of GSR Part 4 (Rev. 1) [15] also stipulates:

"[t]he necessary layers of protection, including physical barriers to confine radioactive material at specific locations, and the necessary supporting administrative controls for achieving defence in depth shall be identified in the safety assessment. This shall include identification of:

(a) Safety functions that must be fulfilled;
(b) Potential challenges to these safety functions;
(c) Mechanisms that give rise to these challenges, and the necessary responses to them;
(d) Provisions made to prevent these mechanisms from occurring;
(e) Provisions made to identify or monitor deterioration caused by these mechanisms, if practicable;
(f) Provisions for mitigating the consequences if the safety functions fail."

Safety Reports Series No. 46 (Rev. 1) [16] provides a possible approach to include all aspects of defence in depth, including human factors and organizational matters, so as to perform the assessment of defence in depth according to the requirements of GSR Part 4 (Rev. 1) [15].

Other methodologies, based on an assessment of defence in depth oriented towards either deterministic or probabilistic design, are also used by States.

To provide an accurate projection of a plant's defence in depth, all identified and related shortcomings are to be included in the assessment, including LTO related findings.

4.6. DETERMINATION OF REASONABLY PRACTICABLE SAFETY IMPROVEMENTS

Paragraph 4.47 of SSR-2/2 (Rev. 1) [3] states:

"On the basis of the results of the systematic safety assessment, the operating organization shall implement any necessary corrective actions and reasonably practicable modifications for compliance with applicable standards with the aim of enhancing the safety of the plant by further reducing the likelihood and the potential consequences of accidents."

SSG-25 [1] states:

(a) "PSR provides an effective way to obtain an overall view of actual plant safety and the quality of the safety documentation, and to determine reasonable and practical modifications to ensure safety or improve safety to an appropriate high level" (para. 2.4; see also para. 3.2);

(b) "Where there are negative findings, the global assessment should provide a justification for any improvements that cannot reasonably and practicably be made" (para. 4.26);

(c) "Overall conclusions and safety improvements considered to be reasonable and practicable in accordance with the global assessment should be documented in the final PSR report" (para. 6.12).

Even though the term 'reasonably practicable' is widely used in IAEA publications and in other sets of requirements, an internationally accepted definition has not been determined. The as low as reasonably practicable principle is widely applicable and is embodied in Principle 5 (optimization of protection) of SF-1 [14]. A graded approach related to the degree of practicality required to resolve deviations could be applied, based on the deviations, LTO corrective measures and risk significance. That is, the necessary justification required to allow issues with a lower risk significance to continue could be less challenging to develop than the justification for issues that could result in large releases.

The key means of preventing and mitigating the consequences of accidents is 'defence in depth'. Paragraph 3.31 3.31 of SF-1 [14] states:

"[w]hen properly implemented, defence in depth ensures that no single technical, human or organizational failure could lead to harmful effects and that the combinations of failures that could give rise to significant harmful effects are of very low probability."

4.6.1. Determination of mandatory safety issues

A significant amount of PSR findings could be of such a nature that their implementation and correction does not impose the use of significant resources by the licensee. Typically, deviations related to management, processes and their documentation are examples of such findings.

Implementation of appropriate, but expensive, corrective measures could be strictly required by national legislation. In such cases, the justification of not implementing a corrective measure is not possible, since the as low as reasonably practicable methodology does not apply to mandatory requirements. The licensee's inability to afford the implementation of costly corrective measures is usually not an accepted argument for considering some of the corrective measures as reasonably not practicable.

4.6.2. Development of safety improvements

Several options for possible solutions for the remaining deviations considered by the global assessment could be developed. These could be drawn from the person who identified the deviation, line organization personnel, subject matter experts or the global assessment team. Safety improvements reducing risk arising from more than one deviation are also possible. It is usually not considered adequate to proceed to the following step with only one option. Different approaches useful to complete this task are available:

(a) Consideration of defence in depth (see Section 4.5);
(b) Screening of best available practice;
(c) PSA;
(d) DSA;
(e) Engineering judgement.

4.6.3. Decision making process

Decisions on which safety improvements are recommended for the period of LTO typically consider the actual benefit to safety that will be achieved through the duration of the remaining planned lifetime of the plant (e.g. para. 6.10 of SSG-25 [1]).

The decision making process typically consists of two activities. The first activity involves consideration of a number of options to identify which provides the best safety benefit. For certain safety issues, PSA may be a useful tool to gain an understanding of the relative safety benefits of alternative options so that decisions are well informed.

The second activity involves a justification of situations where no reasonable safety improvement could be identified. For this purpose, a comprehensive justification is necessary to demonstrate that efforts to implement any of the available safety improvements are grossly disproportionate compared with the safety benefits gained by their implementation.

A description of the aspects to be considered in the evaluation of costs and benefits is outside the scope of this publication. Safety Reports Series No. 12 [13] can provide further information on that subject.

4.6.4. Safety assessment

A safety assessment demonstrating the effectiveness of proposed and implemented safety improvements, compliance with safety requirements and validity of the updated licensing basis is an essential part of the process of identifying and implementing safety improvements.

The results of a safety assessment typically confirm the aspects discussed in Sections 4.2–4.6 for the whole period until the next PSR and, if applicable, the entire period considered for LTO.

4.6.5. Determination of residual risk

The implementation of reasonably practicable improvements includes a demonstration that the associated risks have been reduced to as low as reasonably practicable and their value is consistent with targets set for practical elimination of large or early release, if applicable. The demonstration shows that the costs of any additional safety improvements would be grossly disproportionate compared with the safety benefits that would be achieved by their implementation.

4.7. DEVELOPMENT OF AN INTEGRATED IMPLEMENTATION PLAN

As stated in para. 4.13 of SSG-25 [1], preparation of an integrated implementation plan is the final phase of PSR. The integrated implementation plan contains identified safety improvements and the schedule of their implementation (para. 2.18 of SSG-25 [1]). The integrated implementation plan is submitted to and reviewed by the regulatory body (paras 2.18 and 8.20 of SSG-25 [1]). Safety improvements are implemented in accordance with the integrated implementation plan and the schedule agreed with the regulatory body (para. 6.12 of SSG-25 [1]).

The timely implementation of corrective actions, as well as reasonable and practicable safety improvements resulting from the PSR for LTO, is important. In the typical time frame of implementation all deviations have been addressed before the next PSR, but the more significant issues need to be resolved in the short term. The timing of corrective actions related to LTO partly depends on the starting date of the period of LTO, or it may be related to the results of the revalidation of TLAAs. The methodology for determining the timing of corrective actions related to LTO could differ from that for the resolution of PSR deviations (see Section 3.2.1). If the LTO findings are processed within the PSR global assessment, it could be necessary to adjust the procedure for the timing of corrective measures and/or safety improvements.

During the development of an integrated implementation plan, the following aspects are considered:

(a) Safety significance of deviations;
(b) Licence compliance;
(c) Effectiveness of the safety improvement in resolving the deviation;
(d) Results of the safety assessment obtained in the development of safety improvements;
(e) Root and direct causes of deviations;
(f) Combined effects of findings;
(g) Results of the defence in depth assessment;
(h) Interactions between individual deviations and safety improvements;
(i) Interface analysis results;
(j) Configuration management principles;
(k) Ease of implementation of safety improvements;
(l) Time required for the implementation of safety improvements;
(m) Resource availability;
(n) Cost–safety benefit analysis;
(o) Available time for implementation (to next PSR, or the whole period for LTO).

Taking into account all of these aspects, the implementation of safety improvements included in the integrated implementation plan is planned as soon as reasonable and practicable. When there is immediate and significant risk to the health and/or safety of workers or the public, the corrective measures are implemented promptly without deferring them to after the global assessment results are completed (paras 4.21 and 8.34 of SSG-25 [1]).

Further information on formal aspects of the integrated implementation plan can be found in Section 3.2.2.3 of this publication.

4.8. ASSESSMENT OF THE OVERALL LEVEL OF PLANT SAFETY

The PSR for LTO provides an effective way to obtain an overall view of actual plant safety. The total effect of the negative findings (deviations), safety improvements and positive findings (strengths) identified in the PSR are examined using primarily deterministic methods to ensure that the overall level of plant safety is adequate (e.g. paras 2.4 and 6.10 of SSG-25 [1]). PSA can be used to support an overall view of actual plant safety; however, it is recognized that PSA may only be able to represent (quantify) a limited number of findings and safety improvements.

Where the PSR is used to justify LTO, this justification has to be based on objective evidence that an adequate level of plant safety is ensured for the entire period being considered for LTO. The justification typically includes confirmation that:

(a)   The IAEA's fundamental safety principles (SF-1) [14] are met;
(b)   The plant design meets current standards;
(c)   The adequacy of defence in depth is demonstrated;
(d)   The completeness of the scope of ageing management and LTO is demonstrated;
(e)   The ageing management of in-scope SSCs is adequate;
(f)   The DSA is valid and up to date and confirms adequate safety of the design, including required safety margins;
(g)   The PSA shows a balanced design with no weak points in the plant design;
(h)   The hazard analysis ensures adequate protection against external and internal hazards, with due account taken of the plant design, site characteristics, actual condition of in-scope SSCs, current analytical methods, safety standards and knowledge;
(i)   The safety performance does not show negative trends;
(j)   The proposed safety improvements do not result in any significant residual risks;
(k)   The human resources are adequate for the period of LTO;
(l)   The corrective measures and safety improvements are implemented in a timely manner.

Overall conclusions on plant safety and identified safety improvements in the global assessment are documented in the final PSR report (see para. 6.12 of SSG-25 [1]).

5. IMPLEMENTATION OF PROGRAMMES, COMMITMENTS AND IMPROVEMENTS

5.1. INTRODUCTION

This section addresses the set-up of an integrated implementation programme that responds to the commitments identified, listed and agreed with the regulatory body as a result of the assessments included in LTO and PSR programmes. These commitments determine the scope of the implementation programme that needs to be established, as follows:

(a) Meeting safety and security requirements, despite an extended scope of complex implementation actions;
(b) Optimizing the effective and efficient use of resources, budgets and specific tools as much as possible and applying suitable processes (i.e. purchasing), limiting interfaces within the operational and PSR programme organizations;
(c) Considering intermediate milestones according to the required plan agreed with the regulatory body; for example, for IAEA safety reviews.

The objective of this section is to provide additional information to meet specific safety requirements and related recommendations, as follows:

(a) Requirement 16 of SSR-2/2 (Rev. 1) [3] and para. 7.41 of SSG-48 [2] concerning the programme implementing the corrective actions and improvements to ensure safe LTO, in compliance with the requirements of the national regulatory body and national regulations;
(b) Paragraphs 2.9 and 2.10 of SSG-25 [1] concerning the objectives of a PSR;
(c) Paragraphs 3.5, 3.10 and 6.7 of SSG-25 [1] concerning the nature of improvements, being part of PSR and the role of the PSR global assessment in defining the scope for implementation in the decision to approve LTO;
(d) Paragraph 7.5 (figure 8) of SSG-48 [2] and paras 8.1 and 8.2 (figures 1 and 2) of SSG-25 [1] representing similar major steps for both the programme for LTO, in particular for the ageing management of in-scope SSCs, and for the process for PSR.

5.2. PROCESSES FOR IMPLEMENTATION OF LONG TERM OPERATION AND PERIODIC SAFETY REVIEW

Table 6 illustrates the three major process steps for both LTO and PSR programmes. Table 6 also contains the main products to deliver, a typical timeline and the roles of both the operator and the regulatory body. In most cases, the timeline is determined by national legislation. A requirement may be established that the in-scope work needs to be completed by the date of LTO submission. Other factors having a serious impact on the timeline that require consideration or compliance may include the following:

(a) The requirement to conduct an environmental impact assessment for LTO;
(b) The order and delivery times of qualified materials;
(c) Staggered scheduling of identical works on different redundant safety trains to minimize the risk of common cause failures;
(d) Maximization of the use of the scheduled shutdowns for refuelling to perform the work in scope.

The steps given in Table 6 are indicative to allow the operating organization to continue operations when the new operating period begins. These steps need to be implemented with adequate timing to meet the milestones and deadlines for the LTO submission set by the regulatory body.

TABLE 6. EXAMPLE OF STEPS FOR THE PREPARATION, ASSESSMENT AND IMPLEMENTATION OF THE PROGRAMME FOR LONG TERM OPERATION ALONGSIDE KEY PERIODIC SAFETY REVIEW ACTIVITIES[a]

Process step	Operator	Regulatory body
	Preparation	
1	Feasibility study Business case, demonstrating technical and financial feasibility, based on high level internal and external operational experience	Issuing the regulatory framework and requirements, based on national and international references

TABLE 6. EXAMPLE OF STEPS FOR THE PREPARATION,
ASSESSMENT AND IMPLEMENTATION OF THE PROGRAMME FOR
LONG TERM OPERATION ALONGSIDE KEY PERIODIC SAFETY
REVIEW ACTIVITIES[a] (cont.)

Process step	Operator	Regulatory body
2	Development of LTO concepts, methodology and scope, including: • Basic conditions and assumptions • Development of PSR framework to support LTO • Programme structures • Organizational provisions • Schedule • Programme methodology • Risk evaluation • Management of deliverables List of planned deliverables (e.g. studies, reports, databases)	Coordination with the licensee, commenting documents as needed and initiated by the licensee
3	Submission of the methodology and scope report to the regulatory body	Assessment of the methodology and scope report
4	*PSR — Develop PSR basis document (strategy, scope, schedule of deliverables)*	*PSR — Agree strategy, scope, schedule of deliverables*
	Assessment	
5	Complete assessment report, containing the evaluation of the condition of in-scope SSCs (current and expected for LTO)	Evaluation of programme for LTO, review and assessment of safety factor reports
6	Prepare LTO commitments and final action plan Prepare revised business case	Review LTO commitments
7	*PSR — Complete plant walkdowns* *PSR — Complete safety factor reviews* *PSR — Produce global assessment, including implementation plan*	Review and assessment of results and global assessment Approval of final action plan
8	*PSR — Final submission sent to regulatory body*	*Acceptance and closure of PSR process*

TABLE 6. EXAMPLE OF STEPS FOR THE PREPARATION, ASSESSMENT AND IMPLEMENTATION OF THE PROGRAMME FOR LONG TERM OPERATION ALONGSIDE KEY PERIODIC SAFETY REVIEW ACTIVITIES[a] (cont.)

Process step	Operator	Regulatory body
	Implementation[b]	
9	Implementation and reporting of action plan and demonstration of achieved commitment *PSR — Progress actions and provide confirmation of satisfactory progress on the resolution of actions sent to the regulatory body*	Assessment and inspection of action plan implementation *PSR — Accept PSR actions outcome and agree further actions* *PSR — Monitor the completion of agreed actions and commitments*

[a] Key PSR activities included within the table are highlighted as *italic* text.

[b] The implementation of actions and the sentencing of commitments has to be coordinated, where possible, with existing programmes of work (outage schedule, improvement activities) and has to be completed to the timescales agreed between the licensee and the regulatory body. These activities may be required before the end of the PSR cycle or at an agreed point of time into the next cycle.

The provisional and final decision making on continuation or discontinuation of NPP operation according to the process steps is Member State specific and can depend on various factors, both internal and external to the operating organizations.

5.3. MAPPING OF SIMILAR LONG TERM OPERATION AREAS TO THE SAFETY FACTORS IN THE PERIODIC SAFETY REVIEW

The regulatory body is expected to provide timely notice of its requirements regarding PSR, and in particular LTO, as part of LTO preparation (process step 1 in Table 6). Typical focus areas for both LTO and PSR include the following:

(a) Scope setting;
(b) Ageing management;
(c) Preconditions for LTO related to specific plant programmes, management systems and documentation that need to be effective for ageing management;

(d)   Preserving knowledge and competencies;

(e)   One-time inspection and test programmes, as a specific LTO requirement to prove compliance with the design specifications and review of the cumulative effect of plant modifications;

(f)   Design improvements related to specific concerns such as fire protection or equipment qualification.

The PSR is executed in accordance with the recommendations of SSG-25 [1] through an assessment of 14 safety factors, which can be considered as a grouping of evaluation criteria or review areas. If the plant opts for an integrated PSR and LTO approach, it is recommended to align these review areas, considering the safety factors in the PSR as the backbone structure for assessment. If the regulatory body has not defined this integration, it can be done by the operator as soon as possible and proposed to the regulatory body. Each of the LTO areas to be covered, or parts of these, are assigned to a dedicated safety factor based on the safety factor objective and its associated recommendations. Table 7 presents a typical assignment of LTO areas to safety factors.

Finally, the assessment results for the 14 safety factors are consolidated into a global assessment of strengths, weaknesses and opportunities for improvement.

Some Member States add a 15th safety factor, related to radiation impact on operating personnel, and establish the requirements and applicable evaluation criteria.

To optimize and minimize interfaces among safety factors, in combination with the requirements for LTO, it could be beneficial to reassign certain PSR evaluation criteria to other safety factors and the corresponding LTO areas, to align with Table 7.

Ageing assessment for LTO is a comprehensive analysis that considers detailed experience feedback for SSCs based on their actual condition and predicted ageing. Complementary to LTO ageing management, the assessment of safety factors 8 and 9 provides lessons learned on experience feedback through a review of the effectiveness of the process for operating experience and safety performance. The effectiveness of the review can be based on an analysis of a selection of events that have occurred since the previous PSR. The findings on ageing management for LTO and safety factors 8 and 9 need to be cross-checked during the assessment stage (step 2 in Table 6) and documented as part of the integrated management system for the PSR programme.

This is generally applicable to all safety factors. Operating experience feedback is part of the evaluation to be considered to assess the current status of a safety factor compared to the status in the previous PSR. Specific experience

TABLE 7. CORRELATION BETWEEN SAFETY FACTORS AND AREAS
OF THE LONG TERM OPERATION PROGRAMME

Safety factor		Areas of concern for LTO
SF1	Plant design	Design improvement LTO preconditions related to: • Configuration management and modification management • Configuration management of the design basis
SF2	Actual condition of SSCs important to safety	LTO one-time test and inspections Review of current physical status of in-scope SSCs LTO preconditions related to: • Maintenance • Surveillance and monitoring • In-service inspections • Monitoring of chemical regimes • Corrective actions • Obsolescence management Review of compliance of the existing test and inspection programme with the design basis
SF3	Equipment qualification	LTO preconditions related to equipment qualification
SF4	Ageing	Ageing management for LTO LTO preconditions related to management of TLAAs
SF5	Deterministic safety analysis	Review to determine to what extent existing DSA remains valid for the period of LTO
SF6	Probabilistic safety assessment	Not applicable
SF7	Hazard analysis	Review to determine whether adequate protection against internal and external hazards is ensured for LTO
SF8	Safety performance	Lessons learned from internal operating experience and safety performance Storage capacity for radioactive waste and spent fuel

TABLE 7. CORRELATION BETWEEN SAFETY FACTORS AND AREAS OF THE LONG TERM OPERATION PROGRAMME (cont.)

Safety factor		Areas of concern for LTO
SF9	Use of experience from other plants and research findings	Lessons learned from international operating experience
SF10	Organization, the management system and safety culture	LTO knowledge and competence management related to organization, the management system and safety culture LTO preconditions related to the management system (with a focus on ageing management)
SF11	Procedures	LTO preconditions related to updating the safety analysis report and licensing basis documents
SF12	Human factors	Human factors related to LTO knowledge and competence management
SF13	Emergency planning	Not applicable
SF14	Radiological impact on the environment	Not applicable
	Global assessment	

feedback is collected during the safety factor assessments from sources besides the operating experience feedback process itself, by means of, for example:

(a) 'Engineering experience feedback' on design improvements for LTO;
(b) An international benchmark as part of a one time test and inspection programme for LTO;
(c) The condition assessment of in-scope SSCs and use of international references, such as IGALL;
(d) Interviews conducted for knowledge and competence management for LTO;
(e) On-site interviews on the use of procedures.

5.4. IMPLEMENTATION OF THE PROGRAMMES FOR LONG TERM OPERATION AND PERIODIC SAFETY REVIEW

Before commencement of the assessment stage (step 2 in Table 6), a comprehensive evaluation is made of whether the capacity and capability of the operating organization (e.g. human resources, budget) is sufficient to fulfil the commitments for implementation of the PSR project without jeopardizing the operating organization's operational activities, ensuring the nuclear safety and reliability of power generation. In principle, separate project teams, including both internal personnel and contractors, can be set up to adequately perform the assessment.

As different organizations may be involved in the implementation of the PSR project, good coordination among these organizations needs to be ensured. This includes developing a good understanding of each other's commitments and professional backgrounds, and of the interfaces and interactions among organizations and their specific characteristics, and making any adjustments to these interactions necessary for implementation of the programmes for PSR and LTO.

The essential attributes for the coordination of the programmes for LTO and PSR include the following:

(a) The operating organization's structure

The operating organization's structure for the implementation programme is based either on the areas being considered for LTO or on the safety factors for the PSR. The implementation programme delivers the commitments of the operating organization to the regulatory body. Depending on the size of the scope of the implementation programme, these projects can be bundled again in project portfolios or technical disciplines.

The projects covering the overall scope of the implementation programme are ideally managed in an integrated way through the application of a single overarching project methodology and approach, a unique management reporting system and dedicated interfaces with the regulatory body. The following roles support the implementation programme:

— Programme steering committee and programme owner;
— Programme manager;
— Portfolio or project discipline (mechanical, electrical, I&C, civil) managers;
— LTO assessment leads, PSR safety factor leads;

— Project members, operating within a project portfolio, discipline or reporting to the portfolio or project discipline manager.

(b) Programme organization

The implementation of the action plan for the PSR needs to provide adequate staffing and coordination among different organizational entities or external organizations. The PSR action plan describes a single functional organization's procedure, a single functional organization chart and interface management with the operating organizations and other programmes that are fully oriented towards achieving the objectives of the programme.

(c) Members of the core team

The members of the core team of the organization are fully dedicated to the PSR and LTO programmes for a specified period. This team may be joined by people partially assigned to specific tasks or with particular responsibilities. These employees may remain hierarchically under the responsibility of their department but are functionally attached to the implementation organization.

(d) Organizational matrix

The organizational matrix can be characterized by two dimensions:

— A vertical axis of reporting that is composed of, from the top down, a steering committee, programme manager, domain or portfolio lead and project lead.
— A horizontal or transverse axis determined by the origin of the team members or the competence and the integration of this organization into the operating organization of the plant and, if appropriate, into the corporate engineering. This dimension is also essential for the technical supervision of activities and associated deliverables.

(e) Programme support office

An effective programme support office may be needed as part of the programme organization. This provides technical and administrative support and is essential in accomplishing major project processes such as project planning, reporting and integration management.

A steering committee is established for the PSR or the programme for LTO that receives the required delegations and decision making powers (approval of purchasing strategies and orders specific to this project), within the limits of the overall budget of the programme.

The entire programme is carried out in the greatest possible harmony and coherence with the site operating organization, its departments and processes. To ensure that this runs smoothly, several single points of contact (SPOCs) are ideally appointed at programme level to support this relationship. It is preferable practice to avoid creating separate entities within the implementing organization that already exist within the existing plant organization. It is better to expand the operating organization (with funding from the programme) to manage the extra workload. The role of the SPOC is crucial for providing a bidirectional interface between the implementation programme and the SPOC's organization; specifically, the SPOC:

— Keeps the implementing organization in phase with the programme;
— Brings the concerns of their organization to the programme;
— Keeps up to date with the progress that the implementing organization is making in relation to the programme.

The roles and responsibilities of the SPOCs need to be defined (e.g. there may be an SPOC in the project from operation, as well as an SPOC from each maintenance technical discipline).

(f) Problem solving

If problems arise between the programme organization and the existing organization, these need to be solved at as low a level as possible. In some cases, the plant manager or the programme manager may need to arbitrate, depending on the involved parties (i.e. internal and/or external).

Cross-cutting processes and roles

The following cross-cutting processes and roles have to receive the necessary attention within the programme organization:

— Planning and management of integration and interfaces;
— Contract, insurance and claim management;
— Quality and risk management;
— Participation in external IAEA missions.

5.5. RECOMMENDED TOOLS FOR IMPLEMENTATION OF THE PROGRAMMES FOR PERIODIC SAFETY REVIEW AND LONG TERM OPERATION

5.5.1. Integrated implementation plan

The assessment stage (step 5 in Table 6) leads to the delivery of a set of products such as study documents, reports and recordings in databases. These are listed by PSR safety factor and/or by LTO programme area.

A product can contain no, one or more actions, being part of the future implementation programme (step 9 in Table 6). The resourcing requirements of these actions are also estimated so that possible alternative solutions can be compared. Ultimately, the preferred actions are included in the integrated implementation plan and are considered as the basis for LTO commitments to the authorities. The relationships between the review area and the resulting products and actions must also be preserved and managed in the implementation phase.

Actions are assigned to projects, and projects are grouped into disciplines included in the implementation programme. If eligible, the necessary plant modifications are also identified and listed by project. This relationship between the project and the relevant modification files needs to be managed and maintained. There is no need to have a one to one link between the plant modification file and a specific action, as their goals are completely different. The plant modification process needs to be mature and adequately resourced in order to manage the resulting volume of modifications to avoid backlog problems after realization; ideally, it would be a precondition for LTO.

The integrated implementation plan is the tool to manage the scope of the implementation programme. This plan contains two types of actions:

(a) Actions necessary to comply with the commitments for LTO and PSR;
(b) Other actions not associated with commitments but enabling and ensuring the management of SSCs outside the scope of PSR and the LTO programme (e.g. replacement of SSCs not important to safety, security systems).

A realistic schedule for carrying out the listed actions can only be drafted by positioning these committed actions in the broader context of all projects and line work being undertaken by the plant. Additionally, other critical corrective actions and projects may need to be considered. The scheduling of these other actions may have an influence on the PSR and LTO schedule. These actions, which are not related to LTO, need to be identified and listed to be integrated into the schedule. Examples include:

(a) Work resulting from regulations (e.g. inspections of the reactor pressure vessel, pressure testing of the reactor coolant system);
(b) Work on SSCs not important to safety that are out of the scope of LTO (e.g. work on the turbine control system, replacement of non-classified switchboards);
(c) Long term maintenance tasks (e.g. turbine maintenance, intervention in steam generators, replacement of rotating strainers, transformers).

Only a schedule that integrates all projects and actions will enable a correct assessment of the impact of LTO on nuclear safety, preparation time, workload, priorities and duration of revisions.

A global analysis of these activities and a consolidation of available data will enable the confirmation of the feasibility of possible scenarios and verification of whether applicable preconditions are met.

The integrated implementation plan is the tool that ensures the overview of these objectives and their interrelationships, facilitating the achievement of any commitments to the regulatory body. It can facilitate overall preliminary budgeting and scheduling before the start of detailed project planning and support the overall programme follow-up by adding and completing different types of data, such as the following:

(a) Origin of the improvement (e.g. area, safety factor in the PSR).
(b) Main technical discipline.
(c) Budget estimation with status estimation (e.g. guess, preliminary, detailed).
(d) Human resources estimation.
(e) Earliest and latest dates for start and completion.
(f) Type of work.
(g) Outage time needs or needs for specific operation conditions.
(h) Duration of a few standard project phases such as:
    — Preliminary studies and pre-design;
    — Detailed design;
    — Procurement;
    — Preparation;
    — Realization outside and/or inside shutdown.
(i) Types of associated risk.

The main link between the integrated implementation plan and outage planning is based on the proposed plant modification files and ageing management related activities, specifying requirements for outage planning. These are translated into work orders and grouped into the appropriate phases of outage, fulfilling the respective conditions according to the tasks to be performed.

5.5.2. Programme and project scheduling

The primary objectives of scheduling are the following:

(a) Safety is always ensured, especially in a complex programme that requires numerous interventions.
(b) Technical specifications are always respected, and violations are avoided. If an exception is unavoidable, it is justified and counted. An integrated implementation plan is therefore an essential safety tool, as its logical structure enables an analysis of the safety aspects.
(c) The schedule allows for an appropriate response to any requirement expressed by the regulatory body regarding compliance with the required schedule.

The secondary objectives are the following:

(a) An overview of all aspects of an action (including studies, manufacturing, purchasing and preparations) enables potential bottlenecks (e.g. supplier qualification, manufacturing of components) to be identified beforehand so that contingency planning can be ready in time.
(b) Adequate scheduling of work enables compliance with as low as reasonably practicable obligations. Potential interferences among types of work can be detected immediately so that supplementary analyses with regard to nuclear and industrial safety can be performed as soon as possible.
(c) The extended shutdowns of the NPP (if necessary) need to be determined and integrated into the long term shutdown planning of the NPP to secure the stability of the electricity grid and production capacity.
(d) Demonstration of compliance with the business case, especially regarding production losses.

The programme scheduling contains a multilevel approach to provide the following:

(a) High level management and status reporting;
(b) Medium level scheduling managing time slots for projects, to be used as a living programme instrument;

(c)   Low level scheduling, clarifying the status of actions and commitments, to be used as a living programme instrument.

The overall planning concept needs to be defined considering the respective constraints and assumptions, as well as the dependencies and influences on other NPP activities and programmes.

The following aspects are critical for appropriate planning and need to be considered from the beginning:

(a)   Which actions need to be finalized before PSR or LTO submission, or can be accomplished within an agreed time frame after entering into LTO?
(b)   Which actions demand work inside and/or outside shutdown?
(c)   Which actions demand work that can be accomplished under special operation conditions, such as complete reactor unloading?
(d)   Which actions have a long lead time (e.g. for procurement or qualification) and are decisive for achieving the major milestones in the programme schedule?

A planning team needs to be created to efficiently add expertise in a specific context and eventually recommend an implementation strategy for a specific project. The involvement of experienced operation personnel can be necessary to ensure efficient planning of outages and compliance with technical specifications in connection with estimates or preconditions such as delivery times for spare parts. Organizing work groups with technical expertise, particularly electrical and I&C experts, can be useful to estimate procurement lead times for procurement of qualified materials.

5.5.3. Programme and project collaboration platform

It is useful to set up a collaboration platform that collects information and data from different sources (different domains, teams, etc.) into a single location and provides the team members with easy access to relevant information and products they will require. This tool serves the temporary purposes of the programme and the related projects and its processes. It is not a substitute for the actual plant systems and tools, such as the document management system of the plant. The collaboration platform provides the following:

(a)   A workspace with large storage capacity;
(b)   Libraries for the different types of documentation;
(c)   Workflows to support some project processes;

(d)   Programme specific authorizations for accessibility and user profiles and rights;

(e)   Lists with multiuser functionality and record version management;

(f)   Interaction functions with external parties.

6. ROLES AND RESPONSIBILITIES IN PREPARATION FOR SAFE LONG TERM OPERATION WHEN INTEGRATED WITH PERIODIC SAFETY REVIEW

6.1. GENERAL

This section outlines the roles and responsibilities of all parties involved in the preparation of an NPP for safe LTO. These include the regulatory authority, the NPP teams responsible for the LTO programme and conducting the PSR, and supporting external organizations.

Paragraph 7.1 of SSG-25 [1] and paras 7.4, 7.39 and 7.40 of SSG-48 [2] recommend that the operating organization report all safety significant findings from the PSR to the regulatory body, subject to national regulations. To ensure the safe LTO of an NPP, the operating organization assesses and ensures, while the regulatory body oversees, that the safety of the NPP is maintained throughout the period of LTO in accordance with current safety standards and national regulatory requirements. A PSR used to support LTO includes justification based on trends of expected ageing effects during the period of LTO, informed by past studies. This justification could include studies undertaken in previous PSRs and, when appropriate, plant modifications that have been, or will be, implemented to improve safety.

6.2. ROLES AND RESPONSIBILITIES OF INVOLVED PARTIES

6.2.1. Roles and responsibilities of the regulatory body

As with a typical PSR, the regulatory body has the role of establishing the requirements and any supporting guidance regarding the conducting of the PSR. When a PSR project is begun, a PSR project manager within the regulatory body is appointed, and a PSR regulatory review plan is developed and approved with the assigned resources, in accordance with paras 8.27–8.36 of SSG-25 [1]. The regulatory review plan is typically sent to the licensee for information.

The regulatory body establishes the PSR requirements for LTO, including how PSR outputs can be used for justification of LTO. It is important that these requirements are established well in advance of the planned start of LTO, such that the operating organization knows ahead of time the extent of the work involved in preparation for, and implementation of, LTO. This includes the necessary notification periods required to notify the regulatory body of the intention of engaging in LTO, as well as the regulatory review periods that will be required for key submissions. The regulatory body needs to establish the period within which the documentation is expected to be submitted before reaching the end of the design life.

The regulatory body may consider using the same resources (e.g. internal staff, technical support organization, external experts) to evaluate the submissions for PSR and LTO, in order to improve the efficiency of the regulatory review.

During the development and implementation of the safety factor review, the regulatory body may inspect the implementation of the particular activities performed by the licensee according to the approved regulatory review plan. The regulatory body is typically available for coordination meetings with the licensee before agreement on the PSR basis document as well as during its implementation if any deviation from the agreed activities, schedules and so on is needed.

On completion of the PSR, the regulatory body reviews the results and eventually accepts or approves the results and conclusions, including the proposed safety improvements. In the case of LTO, the regulatory body ensures that the PSR provides an adequate argument to support LTO.

6.2.2. Roles and responsibilities of the operating organization

The operating organization is responsible for carrying out the PSR for justification of LTO, in accordance with regulatory requirements. To discharge this responsibility, the operating organization will have to engage in the following activities:

(a) Prepare for LTO project activities;
(b) Appoint a project management team that includes senior management at the NPP;
(c) Establish project controls (e.g. quality assurance, documentation control, verification and approval plans);
(d) Determine an overall budget and investment profile;
(e) Establish a timeline with significant milestones agreed with all interested parties (operator, strategic partners or contractors, regulatory authorities);

(f) Form a resourcing plan identifying the delivery team, the expected schedule of work and the required competencies, noting that specialist knowledge will be required;

(g) Develop and document strategies, procedures and work instructions to facilitate LTO activities;

(h) Enable LTO activities (e.g. reviews, assessments, walkdowns) to be completed by key staff by either displacing work or adding resources to meet the programme;

(i) Implement improvements to AMPs;

(j) Resource (funding, people, tools, plant availability) LTO tasks, including plant modifications, to an agreed programme;

(k) Ensure documentation management and records control.

PSR and LTO teams in the operating organization are responsible for carrying out the work required for justification of LTO. They determine which information from the PSR can be used for justification of LTO, following the requirements established by the regulatory body. The operating organization is responsible for ensuring that all relevant information provided by the PSR will be made available for justification of LTO.

6.2.3. Desired inputs and outputs of each interested party

The teams involved in PSR and LTO assessments benefit from each other's input to clearly demonstrate the safety of the extended operation of the plant. The operating organization submits all required documentation to the regulatory body so that it can make a decision. The relevant documents include, but are not limited to:

(a) Latest PSR with associated action plan;

(b) Updated FSAR;

(c) Updated DSA and PSA;

(d) AMPs revised for LTO;

(e) Updated technical specifications;

(f) Relevant plant programmes credited for LTO;

(g) Guidelines for severe accident management;

(h) Emergency preparedness and response plan;

(i) Fire protection programme.

The regulatory body communicates the conclusions of the review to the operating organization, along with any conditions as deemed appropriate and in accordance with national requirements. It also communicates the decision to the public.

6.3. COMMUNICATION BETWEEN THE OPERATING ORGANIZATION AND THE REGULATORY BODY

The operating organization maintains regular communications with the regulatory body to ensure that regulatory expectations are clear and to deal with any issues as they arise. A protocol may be established between the operating organization and the regulatory body. The document sets the ground rules about, for example, the activities and their schedules, the representatives on each side and the processes for the resolution of issues. It is signed by senior management representatives on both sides.

Timely interaction between the regulatory body and the operator through notification of intermediate results is an essential factor for the success of both the PSR and LTO programmes. It is advisable that there is periodic and structured consultation with the regulatory body, with the following objectives:

(a) To make clear agreements with regard to the exact scope of the various projects and the expectations (and implicit related commitments and actions) of the regulatory body that must be fulfilled for the projects to be closed in accordance with these agreements;

(b) To discuss evolutions and adjustments in scope and time in a transparent manner so that later setbacks can be avoided;

(c) To discuss the appropriate timing of the resolution of emerging issues, especially those aspects that could have an impact on the acceptance of LTO or PSR by the regulatory body;

(d) To formally close various actions in the integrated implementation plan after the realization of a project.

6.4. COORDINATION OF SUBMISSIONS FOR LONG TERM OPERATION

In an application for LTO, the operating organization is required to demonstrate to the regulatory body that the plant is safe to operate for the period of LTO. The operating organization also needs to demonstrate that the work required to extend the life of the plant, as determined by a PSR that supports the extension of operation, is sufficient to ensure the safety of the plant for the period of LTO.

Justification for LTO may be required to be accompanied by a request for permission to operate beyond an established time frame (e.g. licence renewal, extension of operating permit), depending on national regulations. In this case, the operating organization is responsible for providing, on top of the documentation required for LTO, any additional information required to demonstrate that the plant is safe to operate for the period requested for the licence extension, which may or may not be the same as that for LTO. This may include information on past and current safety performance based on compliance activities carried out by the regulatory body (inspections, document reviews).

Depending on national regulatory requirements, documentation that needs to be provided in a request for LTO includes:

(a) Documentation on programmes related to LTO. The information may be covered in a separate section of the request for LTO.

(b) The updated or new environmental impact assessment (when required by national regulations).

6.5. GOVERNMENT AND PUBLIC INFORMATION

In some Member States, the licensing process for a nuclear facility is public. All documentation submitted as part of the licensing process is publicly accessible, and public hearings are held. Another way of reaching the public is to hold information events in the communities near the facility, to explain the current and proposed work at the facility. This aims to reassure the public that the plant will continue to operate safely in the LTO stage.

6.6. REGULATORY EXPECTATIONS FOR PSR AND LONG TERM OPERATION

The documentation for PSR needs to provide evidence that the safety of the NPP will be maintained throughout the period of LTO and that the licensing basis will remain valid for the period of LTO.

In the LTO approach, the Member State's legislation or the regulatory body may require that safety improvements are in place before entering LTO. Alternatively, a schedule for implementing corrective actions can be agreed with the regulatory body. A series of preconditions needs to be met (e.g. for preserving equipment qualification), and AMPs need to be implemented effectively in order to maintain the safety level of NPPs at the start of the period of LTO.

A programme of inspections and additional tests, if required, can also allow for the verification of the facilities' compliance and the detection of possible issues.

The regulatory body may require adopting a series of minimum regulatory design requirements that require plant modification for LTO. The operator defines and implements the necessary improvements to meet these requirements, as well as the improvements resulting from its own analysis, in order to move closer to the new safety standards and norms. This can be performed independently or in an integrated way. This integration refers to the following combined aspects for PSR and LTO:

(a)   An integrated regulatory framework;
(b)   An integrated methodology and approach, applying one scope definition;
(c)   An integrated assessment report, according to the 14 safety factors in PSR, enriched with the specific requirements for LTO;
(d)   An integrated list of commitments and a list of implementation actions (possibly containing the full scope life extension measures, ensuring reliability and availability for electrical power generation);
(e)   An integrated implementation programme and project approach;
(f)   An integrated final overall review report, serving the objectives of both PSR and LTO.

Further information on regulatory expectations for LTO and for a PSR supporting LTO decision making can be found in Safety Report Series No. 109 [17].

7. DOCUMENTATION IN SUPPORT OF JUSTIFICATION OF LONG TERM OPERATION

This section provides a recommended set of documentation that is developed in the framework of the programmes for LTO and PSR, and a suitable division of the documentation (e.g. results of scope setting, AMR, AMPs, revalidation of TLAAs, review of plant programmes, safety improvements in design) to efficiently justify the safety of LTO. Examples of such documentation include the following:

(a)   Methodology and scope setting of SSCs for assessment of LTO;
(b)   Demonstration that ageing effects are being managed for the period of LTO (AMR, AMPs);

(c) Demonstration that TLAAs have been revalidated for LTO;
(d) Implementation programme for LTO, containing commitments on ageing management and obsolescence management, and an integrated plan for safety improvements identified in the PSR;
(e) Update of FSAR and other current licensing basis documents;
(f) Inputs to licensing documentation from the programmes for LTO and PSR.

Paragraph 4.53 of SSR-2/2 (Rev. 1) [3] states:

"The justification for long term operation shall be prepared on the basis of the results of a safety assessment, with due consideration of the ageing of structures, systems and components. The justification for long term operation shall utilize the results of periodic safety review and shall be submitted to the regulatory body, as required, for approval on the basis of an analysis of the ageing management programme, to ensure the safety of the plant throughout its extended operating lifetime."

Paragraphs 7.29–7.38 of SSG-48 [2] provide recommendations regarding the purpose, content and structure of documentation in support of LTO. This documentation is typically developed as a part of an NPP's LTO programme. Section 6 of Safety Reports Series No. 106 [5] specifies the documentation of ageing management and LTO, including the following:

(a) Documentation of plant level AMP;
(b) Safety analysis report and other licensing documentation;
(c) Documentation of plant programmes for ageing management;
(d) Documentation of the programme for LTO;
(e) Documentation of the scope setting methodology and results;
(f) Documentation of AMR methodology and results;
(g) Documentation of AMPs;
(h) Documentation of ageing management implementation;
(i) Documentation of identification and revalidation of TLAAs for LTO.

In addition, appendix II of SSG-25 [1] provides a recommended scope of documentation for the PSR programme. It describes the content of the documentation of the four main outputs of a PSR as follows (para. II.1):

"The following documents should be produced during the conduct of the PSR to provide the information required by different stages of the process described in this Safety Guide:

— The basis document for the PSR;
— Safety factor report(s);
— The global assessment report;
— The final PSR report, including the integrated implementation plan."

From the perspective of justification of LTO, the most important PSR documents are the global assessment report and the final PSR report. Paragraph 7.37 of SSG-48 [2] states that "[t]he assumptions, activities, evaluations, assessments and results of the plant programme for long term operation should also be reflected in the periodic safety review report, if applicable." Relevant inputs from the LTO programme to the PSR programme are described in Section 3.1 of this publication.

Outputs from the PSR programme that form part of the justification of LTO are described in detail in Section 3.2 and include the following:

(a) Identification of safety upgrades;
(b) Cumulative aspects of ageing and obsolescence;
(c) Conceptual aspects of obsolescence;
(d) Environmental impact of LTO;
(e) Implementation programme of commitments and improvements for LTO.

REFERENCES

[1] INTERNATIONAL ATOMIC ENERGY AGENCY, Periodic Safety Review for Nuclear Power Plants, IAEA Safety Standards Series No. SSG 25, IAEA, Vienna (2013).

[2] INTERNATIONAL ATOMIC ENERGY AGENCY, Ageing Management and Development of a Programme for Long Term Operation of Nuclear Power Plants, IAEA Safety Standards Series No. SSG 48, IAEA, Vienna (2018).

[3] INTERNATIONAL ATOMIC ENERGY AGENCY, Safety of Nuclear Power Plants: Commissioning and Operation, IAEA Safety Standards Series No. SSR 2/2 (Rev. 1), IAEA, Vienna (2016).

[4] INTERNATIONAL ATOMIC ENERGY AGENCY, IAEA Safety Glossary: Terminology Used in Nuclear Safety and Radiation Protection, 2018 Edition, IAEA, Vienna (2019).

[5] INTERNATIONAL ATOMIC ENERGY AGENCY, Ageing Management and Long Term Operation of Nuclear Power Plants: Data Management, Scope Setting, Plant Programmes and Documentation, Safety Reports Series No. 106, IAEA, Vienna (2022).

[6] INTERNATIONAL ATOMIC ENERGY AGENCY, Ageing Management for Nuclear Power Plants: International Generic Ageing Lessons Learned (IGALL), Safety Reports Series No. 82 (Rev. 1), IAEA, Vienna (2020).

[7] UNITED STATES NUCLEAR REGULATORY COMMISSION, Generic Aging Lessons Learned for Subsequent License Renewal (GALL SLR) Report, Final Report, NUREG 2191, Vol. 2, NRC, Washington, DC (2017).

[8] INTERNATIONAL ATOMIC ENERGY AGENCY, Equipment Qualification for Nuclear Installations, IAEA Safety Standards Series No. SSG 69, IAEA, Vienna (2021).

[9] INTERNATIONAL ATOMIC ENERGY AGENCY, Safety of Nuclear Power Plants: Design, IAEA Safety Standards Series No. SSR 2/1 (Rev. 1), IAEA, Vienna (2016).

[10] INTERNATIONAL ATOMIC ENERGY AGENCY, Handbook on Ageing Management for Nuclear Power Plants, IAEA Nuclear Energy Series No. NP T 3.24, IAEA, Vienna (2017).

[11] INTERNATIONAL ATOMIC ENERGY AGENCY, Leadership and Management for Safety, IAEA Safety Standards Series No. GSR Part 2, IAEA, Vienna (2016).

[12] INTERNATIONAL ATOMIC ENERGY AGENCY, Application of the Management System for Facilities and Activities, IAEA Safety Standards Series No. GS G 3.1, IAEA, Vienna (2006).

[13] INTERNATIONAL ATOMIC ENERGY AGENCY, Evaluation of the Safety of Operating Nuclear Power Plants Built to Earlier Standards — A Common Basis for Judgement, Safety Reports Series No. 12, IAEA, Vienna (1998).

[14]   EUROPEAN ATOMIC ENERGY COMMUNITY, FOOD AND AGRICULTURE ORGANIZATION OF THE UNITED NATIONS, INTERNATIONAL ATOMIC ENERGY AGENCY, INTERNATIONAL LABOUR ORGANIZATION, INTERNATIONAL MARITIME ORGANIZATION, OECD NUCLEAR ENERGY AGENCY, PAN AMERICAN HEALTH ORGANIZATION, UNITED NATIONS ENVIRONMENT PROGRAMME, WORLD HEALTH ORGANIZATION, Fundamental Safety Principles, IAEA Safety Standards Series No. SF 1, IAEA, Vienna (2006).

[15]   INTERNATIONAL ATOMIC ENERGY AGENCY, Safety Assessment for Facilities and Activities, IAEA Safety Standards Series No. GSR Part 4 (Rev. 1), IAEA, Vienna (2016).

[16]   INTERNATIONAL ATOMIC ENERGY AGENCY, Assessment of Defence in Depth for Nuclear Power Plants, Safety Reports Series No. 46 (Rev. 1), IAEA, Vienna (in preparation).

[17]   INTERNATIONAL ATOMIC ENERGY AGENCY, Regulatory Oversight of Ageing Management and Long Term Operation Programmes of Nuclear Power Plants, Safety Reports Series No. 109, IAEA, Vienna (2022).

Annex I

MEMBER STATES' EXPERIENCES AND PRACTICES

This section describes the practices various Member States employ when using periodic safety review (PSR) in support of justification of long term operation (LTO).

I–1. UNITED KINGDOM'S PERSPECTIVE AND APPROACH TAKEN

I–1.1. Commercial reactors operating in the United Kingdom

All commercial nuclear reactors currently generating electricity in the United Kingdom (UK) are owned and operated by EDF Energy. The fleet consists of 15 reactors across eight sites. Fourteen of the reactors are advanced gas cooled reactors (AGRs) that were commissioned between 1976 and 1988, having an original financial accounting life of 25 years. The 15th reactor, Sizewell B — a pressurized water reactor that began operation in 1995 — has yet to reach its design life of 40 years.

I–1.2. Regulatory requirements

UK law mandates that the nuclear industry must comply with the Nuclear Installations Act 1965, which requires the licensing of sites used for the installation, operation and decommissioning of nuclear facilities. The licence is a legal document and is granted by the UK's regulatory body, the Office for Nuclear Regulation.

Each site licence contains site specific information, such as the number and type of nuclear installations permitted, as well as a set of 36 standard licence conditions covering design, construction, operation and decommissioning. These conditions are generally non-prescriptive and set goals with which the licensee must demonstrate compliance and adherence to recognized 'good practices'.

Regulatory permission is granted at three levels. These are:

(1)   All licence conditions: Continuous monitoring and/or enforcement.
(2)   Licence condition 30: Permission to restart following a periodic shutdown (refuelling or statutory outages).

(3) Licence condition 15: PSR. Ten-yearly review to assess the current safety case and conformance with modern standards, and identify safety enhancements.

In the UK, the nuclear site licence and the operations on the site are not time limited. There is no formal regulatory requirement for LTO submissions to be provided to the regulatory body. Consent must be given by the Nuclear Decommissioning Authority because of the impact on the Nuclear Liabilities Fund. There is no formal link mandating the use of any data from the existing PSR to support LTO. Changes to the current accounting life can either be for a fixed period (e.g. five years) or based upon best estimates of the currently achievable operating life.

I–1.3. Limitations to the cooperation between the completion of a PSR and assessments for long term operation

Site licence condition 14 requires the licensee to make and implement adequate arrangements for the production and assessment of safety cases consisting of documentation to justify safety during operation. Safety cases for individual SSCs have a range of time validities; some expire prior to the next PSR or before the LTO date, whereas others remain valid beyond the next PSR or LTO date. In all cases, updates to extend are enacted in a timely fashion prior to expiry. There is a requirement to review the safety cases as part of the scope of the PSR either with a detailed review, when dates align, or via the safety case health review process for an SSC. As a result of there being no single point in time when time limited safety cases expire and no requirement to update all these safety cases as part of PSR, there are no absolute synergies among projects.

Decisions made regarding the extension of the current accounting life of the plant and the cessation of power generation remain commercial decisions solely for the licensee to determine. Assessments made to inform this decision focus on lifetime limiting safety cases that could challenge confidence in future safe, financially viable operation (e.g. failure of graphite cores or boilers — irreplaceable SSCs in AGRs). Thorough life management strategies are produced for most systems. These identify ageing and degradation mechanisms and the expenditure required for the extended life of the station. This expenditure is then fed into the business plan. Detailed ageing assessments are completed for relatively few SSCs in support of any decision to pursue LTO. The scope of a PSR is broader, covering additional safety factors, and includes a global assessment that cannot be directly aligned with the summary of assessments carried out for LTO.

In addition, for the fleet of AGRs, the end of the current accounting life and the 10 year interval between PSRs do not always align.

I–1.4. Approach taken and synergies realized

Each of the UK's AGR sites, when considering extending the current accounting life into LTO, produced a lifetime safety review for internal decision makers and a briefing document shared with the Office for Nuclear Regulation. The lifetime safety review supplements other documents (technical reports, financial assessments, supply chain reviews), prepared as part of the LTO assessment, and has the specific goal of identifying potential issues that may be raised by a PSR. The lifetime safety review is based on IAEA guidance for PSR (14 safety factors as identified in IAEA Safety Standards Series No. SSG-25, Periodic Safety Review for Nuclear Power Plants [I–1] and two additional safety factors — radiological protection and decommissioning) and is supported by a statement against each of the IAEA safety factors. There are several safety factors that have a stronger focus on plant ageing and are therefore covered in detail. These are the safety factors associated with the actual condition of SSCs important to safety (safety factor 2), ageing (safety factor 4) and hazard analysis (safety factor 7).

Several work streams required to support LTO and PSR have been undertaken by separate project teams, but the same detailed technical work is often drawn upon by both projects. These teams comprise the station, central technical support, the internal independent regulatory body and external contractor personnel. Effective goal and knowledge sharing between teams working on areas supporting LTO and PSR has resulted in efficiencies in terms of time and cost, and has meant that a significant portion of the work can be completed internally. It has also, when LTO precedes the PSR, provided early warnings of significant technical challenges that will require understanding and resolution as part of the PSR.

Although no direct alignment between the activities performed for LTO and PSR exists, mutual benefit has been realized by both projects at all levels of the organization (executive steering boards, departmental heads and individuals working on topic areas) and by the Office for Nuclear Regulation.

I–2. THE NETHERLANDS' PERSPECTIVE AND APPROACH TAKEN

Borssele nuclear power plant (NPP) (Kerncentrale Borssele) is the only operating nuclear power plant in the Netherlands. This plant underwent an LTO assessment in January 2014 upon reaching 40 years of operation, in order to extend the operating life by 20 years. Dutch nuclear regulation does not contain any specific rules, guidelines or standards pertaining to LTO. At the time of the LTO assessment of Borssele, Safety Reports Series No. 57[1] and IAEA Safety Standards Series No. NS-G-2.12[2], both of which are now superseded by IAEA Safety Standards Series No. SSG-48, Ageing Management and Development of a Programme for Long Term Operation of Nuclear Power Plants [I–2], served as guidelines for the evaluation.

The extent to which PSR and the LTO of the Borssele NPP could benefit from each other's results is discussed in the remainder of this annex.

I–2.1. Long term operation in support of PSR

The third PSR of the Borssele NPP was carried out in the period 2010–2013. In the Netherlands, PSR follows the approach in SSG-25 [I–1]. At the time of the PSR at Borssele, however, SSG-25 [I–1] had not been yet issued; thus, the draft IAEA Safety Guide DS426 was followed instead. Since the LTO assessment had started before the third PSR of the plant, the latter could benefit from several results stemming from the evaluation of LTO. Namely, the PSR relied on the findings of LTO for the review of safety factors 2, 3 and 4. The extent to which the scope of these safety factors was covered by the results of LTO is shown in Fig. I–1.

I–2.2. PSR in support of long term operation

Owing to the fact that the LTO assessment started before the third PSR of the plant, most of the support was provided by the assessment of LTO towards the PSR. Nevertheless, the review of the PSR's safety factors 10 (organization, the management system and safety culture) and 12 (human factors) was part of the LTO licence demonstration as required by the Dutch regulatory body, whereby organizational and administrative aspects, as well as the management system, safety culture and human factors could be reviewed [I–3]. Safety factors 10 and 12, relevant for an LTO licence application, were performed by including the OSART review module 'Management, organization and administration' in the scope of the 2012 SALTO peer review at Borssele.

---

[1] INTERNATIONAL ATOMIC ENERGY AGENCY, Safe Long Term Operation of Nuclear Power Plants, Safety Reports Series No. 57, IAEA, Vienna (2008).

[2] INTERNATIONAL ATOMIC ENERGY AGENCY, Ageing Management for Nuclear Power Plants, IAEA Safety Standards Series No. NS-G-2.12, IAEA, Vienna (2009).

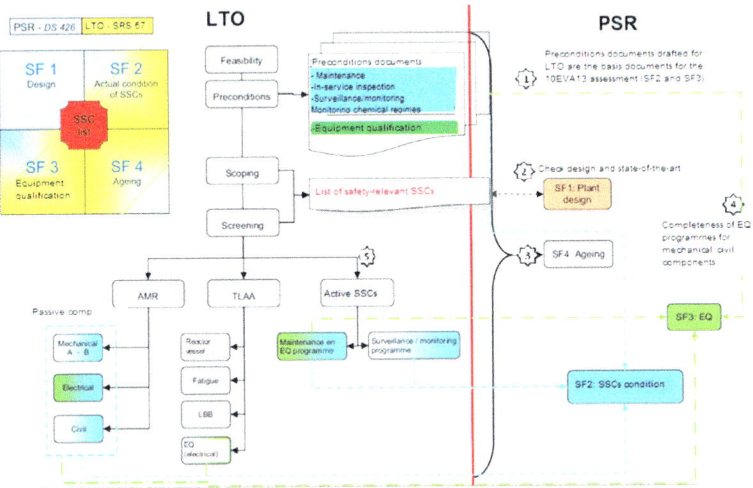

FIG. I–1. Input provided by the long term operation assessment for the third periodic safety review of Borssele nuclear power plant. The grey and yellow box in the top left-hand corner shows qualitatively the degree of overlap between safety factors 1 to 4 and the long term operation assessment. AMR — ageing management review; EQ — equipment qualification; LBB — leak before break; LTO — long term operation; PSR — periodic safety review; SF — safety factor; SSCs — structures, systems and components; TLAA — time limited ageing analysis.

I–3. CANADA'S PERSPECTIVE AND APPROACH TAKEN

I–3.1. Commercial reactors operating in Canada

A total of 19 power reactors are currently operating in Canada. These are located at four NPP sites, each with a power reactor operating licence issued by the Canadian Nuclear Safety Commission (CNSC). They are located in two provinces (Ontario and New Brunswick) and are operated by three distinct licensees (Ontario Power Generation, Bruce Power and New Brunswick Power). The licensed NPP sites comprise between one and eight power reactors, all of which are of the Canada deuterium uranium (CANDU) design. The operating power reactors started generating electricity between 1971 and 1993. The original design life of a CANDU reactor being 30 years, the majority of NPPs in Canada are currently in their LTO period.

I–3.2. Regulatory requirements

The nuclear regulatory body in Canada, the Canadian Nuclear Safety Commission (CNSC), exists under the Nuclear Safety and Control Act (NSCA). The CNSC uses regulations, licences and regulatory documents (REGDOCs) to regulate the nuclear industry.

In Canada, anyone who wants to work with nuclear materials is required to obtain a licence. The licence specifies a validity period within which the licensee can carry on authorized activities, unless that licence is suspended, amended, revoked or replaced. The licence contains conditions specific to the facility, and the licence owner has to comply with these conditions. Before the licence validity period ends, the licence owner has to apply for a licence renewal in order to keep operating.

The NSCA and its regulations do not include specific requirements related to PSR, but licences can include these requirements instead if applicable to that licensee. Regulatory requirements related to PSRs are included in REGDOC-2.3.3 [I–4], which was developed based on SSG-25 [I–1]. In the past, licensees who wanted to operate an NPP beyond its initial design life had to complete an environmental assessment under the Canadian Environmental Assessment Act, along with a PSR. That requirement for an environmental assessment no longer exists.

I–3.3. PSR, long term operation and the licence renewal process

In the early lifetime of NPPs in Canada, PSRs were not used. CNSC staff ensured the safety of the plants via regular compliance activities, such as inspections and document reviews. The licence renewal process allowed the CNSC to assess a licensee's safety performance before issuing a new licence.

As the NPPs were approaching the end of their initial design life, NPP licensees expressed interest in extending the operating life of their plants for another 30 years. PSRs were introduced to determine the reasonable and practical modifications to be completed for LTO. They also enabled NPP licensees to determine the extent to which plants conformed to modern codes, standards and best practices and to identify any factors that would limit safe LTO. The results of the PSRs allowed NPP licensees to decide whether to go ahead with a life extension project based on financial considerations and the CNSC to obtain the necessary information to ensure that plant operation continued to pose no unreasonable risks to health, safety, security or the environment.

Following the completion of PSRs to determine the scope and assess the feasibility of life extension projects, the CNSC decided to require NPP licensees to keep performing PSRs every ten years, in accordance with international practice. Before PSRs were introduced, the validity period of an NPP licence was five years, but this has been extended to ten years to match the validity period of a PSR. When needed, the licence validity period was adjusted such that currently, the results of a PSR are presented at the same time as a request for a 10 year licence renewal.

The NSCA requires the CNSC to hold a public hearing before deciding on a licence renewal. This is done through a well established process in which CNSC staff, as well as the NPP licensee, initially submit documentation to the Commission. When the licence renewal for an NPP includes a request for LTO, the following information needs to be provided:

- A summary and the results of compliance activities carried out in the last licence validity period;
- A summary of the PSR and a list of corrective actions needed for safe operation in the LTO period, along with the implementation schedule;
- An assessment of the NPP's past safety performance;
- For CNSC staff, a recommendation on whether the licence should be renewed and a draft licence.

In addition to CNSC staff and the NPP licensee, members of the public are allowed to intervene during the hearing process to voice their opinions. When the hearing is complete, the Commission makes a decision based on all the information provided and issues a new licence. The licence may include conditions regarding hold points for which CNSC acceptance is required before restarting the reactor.

An NPP licensee in the period of LTO is required to keep performing a PSR every ten years and presenting the results, along with a licence renewal application, until the end of operation.

I–4. SPAIN'S PERSPECTIVE AND APPROACH TAKEN

I–4.1. Commercial reactors operating in Spain

A total of seven reactors across five sites are currently operating in Spain. They are operated by three distinct licensees (Asociación Nuclear Ascó-Vandellòs II A.I.E., Centrales Nucleares Almaraz-Trillo A.I.E. and Iberdrola Generación S.A.U.). Six of the reactors are pressurized water reactors using US (five) and German (one) technology. One reactor is a boiling water reactor using US technology.

The reference date to consider the beginning of LTO for nuclear reactors in Spain is 40 years from the plant's first connection to the national grid. Operating NPPs in Spain were first connected to the national grid between 1981 and 1988. Besides the required operating licence, which is subject to a period of validity, there is no legal limitation on the operating lifetime of NPPs as long as national safety requirements are met and high plant safety levels are demonstrated for the next operation period.

I–4.2. Regulatory requirements

The regulatory requirements on nuclear facilities are governed by Royal Decree 1836/1999 on the Regulation on Nuclear and Radioactive Facilities (RINR).

The RINR sets out the responsibilities of the Ministry of Industry, Energy and Tourism (now Ministry for the Ecological Transition and Demographic Challenge) for granting the nuclear licences needed for the siting, construction, operation, modification, transport, dismantling and decommissioning of facilities. A licence is granted on condition of receipt of a report from the Spanish Nuclear Safety Council (CSN) on the safety evaluation of the application.

The CSN is, in accordance with Law 15/1980, the sole nuclear safety and radiation protection authority in Spain. The CSN is an entity governed by public law, but it is independent from the central government.

The responsibilities of the CSN, as regulatory body, are established in Law 15/1980 and Royal Decree 1440/2010. They include the continuous oversight of nuclear and radioactive installations to ensure they meet safety criteria, with the final mission of protecting workers, the public and the environment. Among its responsibilities, the CSN provides the Spanish Government with the necessary regulations on nuclear safety and radiological protection, and evaluates licensees' applications requiring authorization in accordance with national regulations.

Operating any nuclear installation requires, in accordance with the RINR and its subordinated regulations, an operating licence issued by the Ministry. Operating licences are time limited and contain limits and conditions specific to the installation. Before the validity period of the operating licence expires, the licensee can apply for a renewal following the requirements established in the current operating licence, the RINR and specific regulations (safety instructions issued by the CSN). Operating licences are typically requested and issued for a validity of ten years, although no time limitation is set in the Spanish regulations.

Operating licence renewals are evaluated by the CSN, which takes into account, among other aspects, the CSN's continuous evaluation through the integrated system for the oversight of NPPs and the PSR conducted by the licensees every ten years.

The concept of PSR was introduced in Spain in the early 1990s, before the first operating licences were issued. In the early life of NPPs, they were authorized to operate under a provisional operating permit issued for less than two years' duration. During the 1990s, the CSN established the approach for the transition from the provisional operating permits to operating licences subject to a period of validity, which introduced the performance of a PSR every ten years to complement continuous evaluation of plant safety.

The first guidelines for PSR were developed in 1995 (GS 1.10 Rev. 0, CSN Safety Guide on Periodic Safety Reviews of Nuclear Power Plants), and described the objectives, scope and methodology for conducting PSR in NPPs. Since the late 1990s, all NPPs in Spain have conducted a PSR every ten years as part of the analysis required in the operating licence renewal process.

Performing a PSR every ten years is a legally binding requirement in accordance with Safety Instruction 26 (IS-26 on Fundamental Nuclear Safety Requirements Applicable to Nuclear Installations).

GS 1.10 (currently in rev. 2) was revised in 2017 with the following objectives:

(a) Adopt IAEA guidelines for conducting PSR (SSG-25) [I–1];
(b) Incorporate the lessons learned from the Fukushima Daiichi NPP accident;
(c) Increase focus on LTO challenges;
(d) Develop WENRA Safety Reference Levels for PSR.

GS 1.10 includes the specific recommendations applicable to PSR that are used to justify safe LTO (last PSR prior to the start of the LTO period), which include the following additional analyses:

(a) Integrated plan for ageing assessment and management (including ageing management review (AMR) for LTO and time limited ageing analysis (TLAAs));
(b) Supplement to the final safety analysis report (FSAR), including LTO considerations;
(c) Technical specification modification proposal to consider LTO aspects;
(d) Analysis of the LTO radiological impact;
(e) Radioactive waste management plan, including LTO considerations.

The first specific regulations for LTO in Spain were issued in 2009 through the legally binding Safety Instruction 22 (IS-22 on Safety Requirements for the Management of Ageing and Long-Term Operation of Nuclear Power Plants).

IS-22, which applies during the design lifetime and during LTO periods, sets the criteria for effective ageing management of plant components and also the additional ageing management requirements for LTO. These LTO requirements include the following:

(a) The development of an integrated plan for ageing assessment and management, which summarizes the technical justification of ageing management related to LTO;

(b) The development and implementation of an ageing management plan (AMP) for LTO, which includes AMPs with the necessary activities to monitor, control and mitigate the ageing and degradation mechanisms identified in the integrated plan for ageing assessment, as well as others that may arise during the period of LTO.

As described earlier, in Spain, LTO and PSR documentation are required and evaluated by the CSN through the decision making processes to renew operating licences beyond the period of LTO.

I–4.3. PSR in Spain

A PSR conducted every ten years complies with the national safety guide GS 1.10, which follows SSG-25 [I–1]. The objectives of the PSR are:

(a) To evaluate the suitability and effectiveness of plant programmes and SSCs to maintain safe operation until the next PSR (or the end of commercial operations, if this occurs before the next PSR);

(b) To verify the level of compliance against the most recent nationally and internationally applicable codes and standards, and best practices regarding safety aspects;

(c) To identify necessary actions to correct any possible deviation from the current licensing basis, which may be identified as a result of the review;

(d) To develop an improvement plan based on the review results (positive and negative findings) to maintain and improve safety levels, ensuring that the plant maintains safety until the next PSR (or the end of commercial operations, if this occurs before the next PSR);

(e) To identify necessary improvements to safety documentation, including the current licensing basis, until the next PSR (or the end of commercial operations, if this occurs before the next PSR).

The project for PSR typically consists of four main stages:

(1) Development of the PSR basis document, which details the scope, objective and methodology of the PSR. The PSR basis document requires acceptance

by the CSN, so that this document is agreed between the regulatory body and the licensee prior to the start of PSR activities. The PSR basis document is submitted to the CSN for acceptance at least 15 months before the PSR submittal date.

(2)   Safety factor review. Sixteen safety factors (14 safety factors identified in SSG-25 [I–1], and two additional safety factors on radiological impact on workers and the public and other plant programmes not covered by other safety factors) are reviewed following the methodology presented in the PSR basis document.

(3)   Global assessment. This presents an integrated evaluation of plant safety levels considering the results from the safety factor review. The global assessment includes a summary of the conclusions of each safety factor review, an analysis of PSR results (interface analysis, safety categorization of PSR findings, identification of safety improvements to address PSR findings, integrated implementation plan for PSR safety improvements) and an overall assessment of plant safety levels for the next PSR period. The global assessment follows the methodology presented in the PSR basis document.

(4)   Evaluation by the regulatory body. PSR documentation, together with other documentation required for operating licence renewal, is evaluated by the CSN. The PSR evaluation is used in support of the decision making process for operating licence renewal and may result in additional improvement actions and/or changes to be incorporated into the integrated implementation plan presented by the licensee. These possible additional improvement actions, which are to be implemented during the next operating licence period, are normally documented as licensee commitments and/or limits and conditions of the new operating licence or complementary technical instructions issued by the CSN.

I–4.4. Long term operation

The programme for LTO in Spain is mainly focused on physical ageing, in accordance with the regulatory requirements described in Section I–4.3.

Lifetime management is a continuous process implemented during the initial design lifetime of the plant, which continues during the period of LTO, integrating all results from the LTO analysis.

The lifetime management programme complies with IS-22, which is based on the United States Nuclear Regulatory Commission (USNRC) approach to licence renewal applications (including 10CFR54, RG 1.188 [I–5], NEI 95-10 [I–6], NUREG-1800 [I–7] and NUREG-1801 [I–8]).

The scope of the lifetime management programme covers all passive and long life SSCs with functions important to safety according to IS-22 requirements. For SSCs identified as within the scope, each potential ageing mechanism and effect is analysed as part of the AMR. Plant activities are evaluated to ensure adequate management, mitigation and control of those significant ageing mechanisms. As a result, the necessary improvements to plant activities are identified, and the activities are integrated into the different AMPs. AMPs are implemented prior to LTO and continuously improved taking into consideration the results of the AMP implementation, review of related operating experience, review of new codes and standards, review of new plant modifications and results from internal and external process evaluations and inspections.

LTO assessments are developed and submitted to the CSN three years before the renewal of the operating licence for operation beyond the initial design lifetime. The various assessments for LTO include the following:

(a)   Integrated plan for ageing assessment and management (including AMR for LTO and TLAAs);
(b)   Supplement to FSAR, including LTO considerations;
(c)   Technical specification modification proposal to consider LTO aspects;
(d)   Analysis of LTO radiological impact;
(e)   Radioactive waste management plan, including LTO considerations.

The lifetime management programme is reviewed by the licensee to integrate the conclusions and improvements identified in the LTO assessments, resulting in the lifetime management programme for LTO. This programme is a continuous process implemented during the period of LTO to ensure the adequate management, mitigation and control of significant ageing mechanisms and to confirm that TLAA considerations are maintained.

The assessments for LTO listed above are supplemented by the PSR analysis conducted every ten years. The PSR has a broader scope that addresses additional LTO related areas involving time related challenges, which cannot be directly aligned with the LTO assessments carried out in accordance with national regulations.

Additional areas related to LTO evaluated during the PSR include the following examples:

(a)   Ageing management of active SSCs (covered by continuous plant processes such as maintenance rule, in-service inspection, and programmes for environmental qualification, equipment reliability and proactive obsolescence management);
(b)   Technological obsolescence;

(c)   Regulations, codes and standards obsolescence;
(d)   Knowledge obsolescence;
(e)   Configuration management.

I–4.5. Coordination and synergies between PSR and long term operation

The LTO programme, which includes all the assessments for LTO required by national regulations, results in the lifetime management programme for LTO, which is continuously implemented at the plant during the period of LTO.

PSR, as a comprehensive periodic assessment, is used as a driver for the improvement of plant safety levels and the plant processes implemented continuously at the plant; however, it is not considered as a replacement of these plant processes.

Since PSR and LTO assessments are required by the CSN in the decision making process for the operating licence renewal, both programmes are normally conducted within a close time frame.

Programmes for LTO and PSR are undertaken by separate licensee project teams, as the two programmes have a very different scope in Spain. The programme for PSR has a broader scope, which covers all areas important to safety (16 safety factors), whereas the programme for LTO is mainly focused on physical ageing management, as described in Section I–4.4.

Although the programmes for LTO and PSR are performed separately, they are aligned in time (through the operating renewal process), and significant synergies are identified, which represents a mutual benefit from the licensee's and the CSN's perspectives.

From the licensee's perspective, to ensure adequate coordination between both programmes, members of the LTO project team are involved in the analysis of safety factor 4 (ageing) of PSR. Also, actions for safety improvement affecting LTO are identified and assigned to the LTO coordinator, in order to integrate these improvements into the lifetime management programme for LTO.

From the regulatory body's perspective, the documentation of the PSR and LTO is evaluated together as part of the evaluation of the operating licence renewal application, which provides a solid justification for safe continued operation, allows sharing of evaluation resources and ensures consistency of documentation packages and their evaluation.

REFERENCES TO ANNEX I

[I–1]   INTERNATIONAL ATOMIC ENERGY AGENCY, Periodic Safety Review for Nuclear Power Plants, IAEA Safety Standards Series No. SSG-25, IAEA, Vienna (2013).

[I–2] INTERNATIONAL ATOMIC ENERGY AGENCY, Ageing Management and Development of a Programme for Long Term Operation of Nuclear Power Plants, IAEA Safety Standards Series No. SSG-48, IAEA, Vienna (2018).

[I–3] ELEKTRICITEITS-PRODUKTIEMAATSCHAPPIJ ZUID-NEDERLAND, Aanpassing Veiligheidsrapport inzake bedrijfsduur Kerncentrale Borssele (Long Term Operation), EPZ (2012).

[I–4] CANADIAN NUCLEAR SAFETY COMMISSION, Periodic Safety Reviews, Regulatory Document Series No. REGDOC-2.3.3, CNSC, Ottawa, Canada (2015).

[I–5] UNITED STATES NUCLEAR REGULATORY COMMISSION, Standard Format and Content for Applications to Renew Nuclear Power Plant Operating Licenses, Regulatory Guide 1.188, Rev. 2, NRC, Washington, DC (2020).

[I–6] NUCLEAR ENERGY INSTITUTE, Industry Guideline for Implementing the Requirements of 10 CFR Part 54 — The License Renewal Rule, NEI 95-10 (Rev. 6), NEI, Washington, DC (2005).

[I–7] UNITED STATES NUCLEAR REGULATORY COMMISSION, Standard Review Plan for Review of License Renewal Applications for Nuclear Power Plants, Final Report, NUREG-1800, Rev. 2, NRC, Washington, DC (2010).

[I–8] UNITED STATES NUCLEAR REGULATORY COMMISSION, Generic Aging Lessons Learned (GALL) Report, Final Report, NUREG-1801, Rev. 2, NRC, Washington, DC (2010).

Annex II

EXAMPLE OF A GLOBAL ASSESSMENT

This section provides Member States' practices and lessons learned when performing global assessment.

II–1.1. CZECH REPUBLIC'S PERSPECTIVE AND APPROACH TAKEN

II–1.1. Legal framework

The Czech Republic issued a new Atomic Act that includes the original Atomic Act and a set of implementing decrees in 2017. The Atomic Act includes all the requirements of European Union legislation with respect to WENRA Safety Reference Levels and the relevant recommendations of the IAEA Safety Guides. The Czech regulatory authority, the State Office for Nuclear Safety (SÚJB), issued a new Decree on Safety Assessment, stipulating that safety assessment shall be used to evaluate relevant information about the risks associated with the use of nuclear energy and to adopt measures to prevent compromising the level of safety.

The Decree on Safety Assessment describes in detail all the requirements for periodic safety review (PSR) performance. It describes the scope and content of areas to be assessed, required documentation (strategy, methodology, reports) and findings evaluation. The PSR evaluates the state of the nuclear installation during operation and decommissioning. As part of the PSR, the safety significance of any identified deviations from the safety requirements is evaluated, including defence in depth.

Following this major update of national legislation, the SÚJB made a significant effort to develop a broad set of national safety guides covering major areas of regulated activities, providing detailed guidance for licensees in implementing legal requirements. The national Safety Guide on Periodic Safety Review, which was originally based on IAEA Safety Standards Series No. SSG-25, Periodic Safety Review for Nuclear Power Plants [II–1], has been significantly updated, taking into account the new nuclear legislation.

The new nuclear legislation, the latest IAEA Safety Standards and the WENRA Safety Reference Levels have been applied in recent PSRs in the Czech Republic. The first PSR with respect to the Decree on Safety Assessment was performed in 2020 for nuclear power plant (NPP) Temelín (within the documentation attached to the application for a permit for operation).

II–1.2. Objectives of the global assessment

The approach to the global assessment within the PSR in the Czech Republic is consistent with guidance provided in SSG-25 [II–1], taking into consideration the specific requirements of the new nuclear legislation and SÚJB guidance.

As stated in SSG-25 [II–1], the objective of the global assessment is to develop comprehensive justification of the continued operation of the nuclear facility based on all the results of safety factor reviews, including identified negative (deviations) and positive (strengths) findings, and proposed corrective measures and safety improvements.

The comprehensive justification of the continued operation is based on an evaluation of the impact of all findings from safety factor reviews on the safety of the nuclear facility. It addresses operation before the implementation of corrective actions and cases where no reasonably applicable safety improvements have been identified.

II–1.3. Activities performed in the global assessment framework

Within the global assessment, the following key activities are performed:

(a) Grading of deviations;
(b) Analysis of interfaces, gaps and overlaps among the different safety factor reviews;
(c) Analysis of the combined effects of deviations;
(d) Assessment of defence in depth;
(e) Development of an integrated implementation plan;
(f) Evaluation of the overall level of plant safety.

Specific information on approaches used by the ČEZ company in the Czech Republic is provided below. The information given represents the status of these activities as of the day of preparation of this text; nevertheless, all activities are subject to continuous improvement processes.

II–1.3.1. Grading of deviations

The grading of deviations (negative findings) is based on their safety significance. Evaluation of the safety significance of deviations identified in safety factor reviews is based on the risk matrix adopted from Safety Reports Series No. 12, Evaluation of the Safety of Operating Nuclear Power Plants Built to Earlier Standards — A Common Basis for Judgement [II–2]. The original matrix has been updated to also cover extremely rare (remote) frequencies, and the safety significance of scenarios with intolerable consequences has been increased by one level for unlikely and remote frequencies. Table II–1 presents the updated matrix.

In depth analysis of each finding is performed by a multidisciplinary team to derive the parameters needed to determine the safety significance of deviations according to the risk matrix. The following steps are included in the analysis process:

(a) Analysis of the impact of the deviation on safety functions;
(b) Determination of fundamental safety functions affected by the deviation;
(c) Identification of accident events and sequences relying on the affected fundamental safety functions;
(d) Determination of accident events and frequencies of accident sequences;
(e) Determination of accident events and radiological consequences of accident sequences.

In cases where no safety function can be identified as affected by the deviation, a special procedure based on impact assessment on processes and activities important to safety is applied. Examples of these cases are deviations related to management systems, human factors and procedures. Deviations related to the fifth layer of defence in depth are also assessed using this special procedure.

The results of the analysis of the safety significance of deviations are used to prioritize corrective measures and safety improvements included in the integrated implementation plan. Consistent with SSG-25 [II–1], deviations ranked as issues having high safety significance are given the highest priority, and their resolution is initiated immediately after the classification is confirmed. When this situation occurs, continued operation of the plant needs to be supported by justification for continued operation. The following categories represent the results of the grading of deviations:

(a) HIGH — deviations require immediate action;
(b) MEDIUM — to be resolved within a short period of time;
(c) LOW — to be resolved by the next PSR;
(d) NEGLIGIBLE — to be resolved within routine working processes (usually by the next PSR);
(e) NO IMPACT — formal findings, not required to be resolved.

An example of a graphical presentation of the grading of deviations is shown in Fig. II–1.

TABLE II–1. AN EXAMPLE OF GRADING DEVIATIONS IDENTIFIED IN PERIODIC SAFETY REVIEW

Potential consequences	Tolerable			Significant			Intolerable		
Safety function capability	Robust	Adequate	Inadequate	Robust	Adequate	Inadequate	Robust	Adequate	Inadequate
Event frequency									
Expected	Low	Low	Medium	Medium	High	High	High	High	High
Possible	Very low	Very Low	Low	Low	Medium	High	Medium	Medium	High
Unlikely	Very Low	Very Low	Very Low	Very Low	Low	Low	Low	Low	Medium
Remote	Very Low	Very Low	Very Low	Very Low	Very Low	Very Low	Very Low	Low	Low

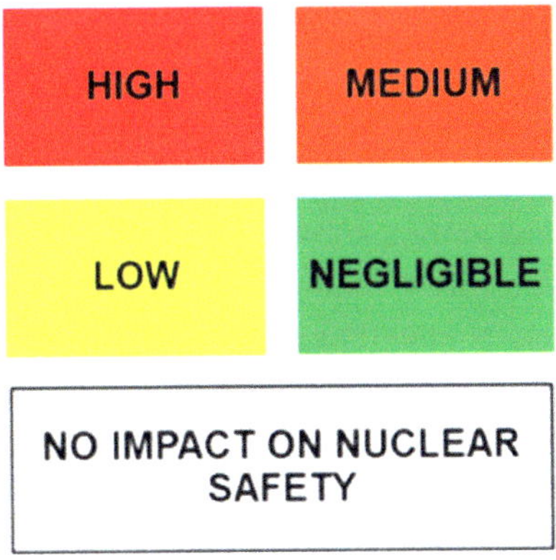

FIG. II–1. An example of a graphical presentation of the grading of deviations.

II–1.3.2. Analysis of interfaces, gaps and overlaps among the different safety factor reviews

The analysis of interfaces, gaps and overlaps is intended to confirm a PSR's completeness and the consistency of its results.

Confirmation of the completeness of the PSR is necessary to provide assurance of meeting its objectives. Completeness of the PSR is ensured by a comprehensive PSR basis document and strict adherence to this document in the review of safety factors.

The operator of the Czech NPPs (Dukovany and Temelín), ČEZ, a. s., invited the IAEA in 2017 to provide a technical safety review service focused on reviewing the consistency of the PSR basis document with the latest IAEA standards. The IAEA team reviewed ČEZ's PSR basis document, which had been completely reworked following a major Czech legislative update. Consistency with the IAEA set of standards has been confirmed, providing assurance that the scope of the PSR basis document provides the fundamental precondition for a comprehensive PSR.

An independent review is a fundamental tool that helps ensure the completeness of the PSR. The independent review of safety factor assessment reports is performed by an independent review team consisting of highly qualified staff and includes experts from technical support organizations. The independent review team is responsible for checking the consistency of the safety factor review with the requirements of the PSR basis document and for confirming the adequacy and correctness of assessments.

The activities described above are cornerstones of the review of completeness of the PSR in the global assessment.

Another task in the global assessment is the confirmation of the review output consistency and analysis of interfaces. Interfaces among safety factors as described in SSG-25 [II–1] represent situations where the results of one safety factor review are used as inputs in another safety factor review. In fact, this situation also happens among sub-elements within one safety factor. These interfaces represent a significant challenge to PSR project management, which led ČEZ, a. s., to pay special attention to the description of interfaces within the PSR basis document and reflect these interfaces in the project plan for the PSR. By this approach, the iterative nature of the review has been minimized, thus saving resources and reducing the time necessary to complete the review.

Within the global assessment, the impact of deviations on the quality of information included in the interface links is checked, and the validity of review results is analysed.

II–1.3.3. Analysis of the combined effects of deviations

Even though each deviation identified in the safety factor review is graded, situations where several deviations have synergic effects need to be considered in the global assessment. In some cases, strengths could be given credit to lower an identified impact of synergic effects of deviations.

Analysis of combined effects is performed by a multidisciplinary assessment team as an iterative process. The analysis process starts with the deviations with the highest safety significance to limit the number of deviations and identify the synergies with the highest impact on safety. In the next runs, deviations with lower safety significance are included in the analysis. The last activity is identifying the possibility of compensating significant combined effects with strengths, if necessary.

There are two types of synergic effects among deviations that could be used to identify the combined effects of deviations with a higher impact on safety than individual deviations:

(a)   Direct synergy;
(b)   System problem indicators.

Direct synergy means deviations causing a degradation of several barriers, leading to a higher probability of initiating events or a decreasing ability of the plant to handle these events. An example could be deviations related to the design of the control room together with deficiencies in control room staff training.

The system problem indicator is usually a larger set of deviations related to the same activity. They could be separately considered as isolated flaws, even with low or negligible safety significance, but all together could indicate a possible system problem with the activity.

The results of the analysis of the combined effects of deviations are used to identify new corrective measures in the integrated implementation plan or to modify the content and/or priorities of existing corrective measures.

II–1.3.4. Assessment of defence in depth

Assessment of defence in depth represents the real challenge for the global assessment because of its complex nature. The result of research carried out by the ČEZ company on possible ways to perform this complex activity was the decision to use the methodology of objective trees described in Safety Reports Series No. 46 (Rev. 1), Assessment of Defence in Depth for Nuclear Power Plants [II–3]. The method is consistent with the requirements established in IAEA Safety Standards Series No. GSR Part 4 (Rev. 1), Safety Assessment for Facilities and Activities [II–4], and provides a complex approach to assessing the robustness of defence in depth layers and their independence.

The detailed technical approach to assess design provisions for defence in depth is analysed by special engineering approaches and tools (i.e. a software tool for the functional analysis of defence in depth) developed by the ČEZ company within the design basis reconstitution project.

For the effective use of the methodology of objective trees, the PSR basis document and the global assessment methodology had to be adapted. Analysis of objective trees showed a strong correlation between provisions preventing mechanisms and challenges from occurring and requirements (criteria) based on the review of safety factors for the PSR, as shown in Fig. II–2.

Adding provisions to relevant criteria in the methodology for the review of safety factors (PSR basis document) was a necessary precondition to complete the implementation of the provisions for screening.

The independent assessment of the implementation of provisions resulting from the safety factor review was performed by the global assessment team to validate the results and to analyse the impact of those provisions that were not implemented on the protection of mechanisms and challenges to fundamental safety functions on each layer of defence in depth.

Finally, an analysis of the mechanisms and challenges affecting the provisions that were not implemented on more than one layer of defence in depth is performed.

The results of the defence in depth assessment are used to confirm the status of defence in depth implementation, help identify its weak points and adjust corrective measures and safety improvement priorities in the integrated implementation plan.

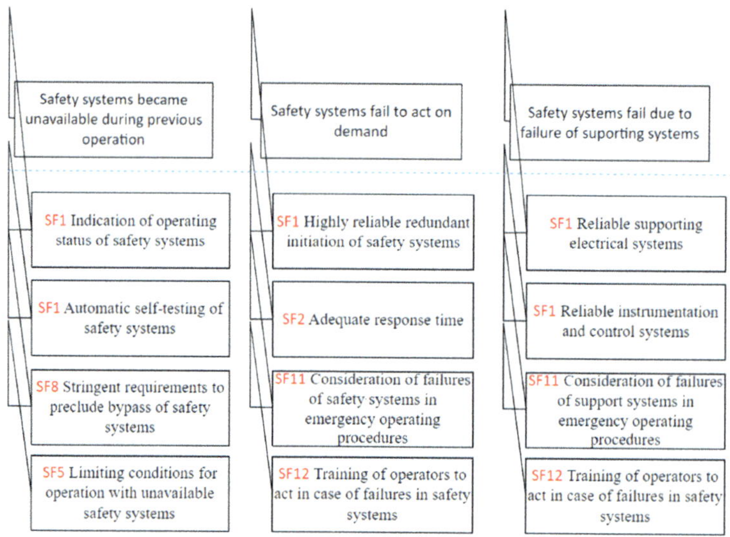

FIG. II–2.   An example of correlations between provisions preventing mechanisms and challenges.

II–1.3.5. Development of an integrated implementation plan

The development of an integrated implementation plan is a complex process, reflecting not only PSR results but also the results of other safety relevant activities and projects related to the licence renewal process, and/or activites related to long term operation (LTO).

Measures resulting from these activities are grouped into complex tasks (aggregate topics) using topical and organizational synergies in the implementation of individual corrective actions and safety improvements. The final integrated implementation plan is approved, and its implementation status is verified twice a year by the high level managerial committee chaired by the plant director.

II–1.3.6. Evaluation of the overall level of plant safety

For the evaluation of overall plant safety and its suitability for continued operation, the results of all safety factor reviews and the results of the global assessment are used. Confirmation of conformance with the newest safety standards to a reasonably applicable extent, confirmation of the SSCs' health, the availability of sufficient numbers of qualified staff, an adequate safety assessment confirming the current design basis and a management system providing sufficient assurance that identified deviations will be resolved in a timely manner constitute the basis for continued operation of the plant.

Based on all this information, the prediction of safe operation for the period until the next PSR is formulated. If the PSR supports LTO, the prediction is made until the end of planned operation.

II–1.4. Results of global assessment

The results of the global assessment are documented in the final report of the PSR, which is provided to the regulatory body together with the complete set of safety factor review reports and an integrated implementation plan. The status of corrective actions and implementation of safety improvements is reported to the regulatory body annually.

REFERENCES TO ANNEX II

[II–1] INTERNATIONAL ATOMIC ENERGY AGENCY, Periodic Safety Review for Nuclear Power Plants, IAEA Safety Standards Series No. SSG-25, IAEA, Vienna (2013).

[II–2] INTERNATIONAL ATOMIC ENERGY AGENCY, Evaluation of the Safety of Operating Nuclear Power Plants Built to Earlier Standards — A Common Basis for Judgement, Safety Reports Series No. 12, IAEA, Vienna (1998).

[II–3] INTERNATIONAL ATOMIC ENERGY AGENCY, Assessment of Defence in Depth for Nuclear Power Plants, Safety Reports Series No. 46 (Rev. 1), IAEA, Vienna (in preparation).

[II–4] INTERNATIONAL ATOMIC ENERGY AGENCY, Safety Assessment for Facilities and Activities, IAEA Safety Standards Series No. GSR Part 4 (Rev. 1), IAEA, Vienna (2016).

ABBREVIATIONS

AGR	advanced gas cooled reactor
AMP	ageing management programme
AMR	ageing management review
CNSC	Canadian Nuclear Safety Commission
CSN	Spanish Nuclear Safety Council
DSA	deterministic safety analysis
FSAR	final safety analysis report
I&C	instrumentation and control
IGALL	international generic ageing lessons learned
LTO	long term operation
NPP	nuclear power plant
PSA	probabilistic safety assessment
PSR	periodic safety review
RINR	Regulation on Nuclear and Radioactive Facilities
SSCs	structures, systems and components
TLAA	time limited ageing analysis
WENRA	Western European Nuclear Regulators' Association

CONTRIBUTORS TO DRAFTING AND REVIEW

Bailey, M.	EDF Energy, United Kingdom
Bertocchi, F.	Nuclear Research and Consultancy Group, Netherlands
Cserhati, A.	Paks II, Hungary
Dang, N.	Canadian Nuclear Safety Commission, Canada
Duchac, A.	International Atomic Energy Agency
Eisan-Kouznetsova, S.	Canadian Nuclear Safety Commission, Canada
Foster, N.	Eskom, South Africa
Gingras, I.	Canadian Nuclear Safety Commission, Canada
Hiser, A.	Nuclear Regulatory Commission, United States of America
Kastelijn, E.	Tractebel, Belgium
Krivanek, R.	International Atomic Energy Agency
Machacek, J.	ČEZ, a.s., Czech Republic
Manojlovic, S.	NEK, Slovenia
Nieto, J.	Centrales Nucleares Almaraz-Trillo, Spain
Padernera, P.	Nucleoeléctrica Argentina S.A., Argentina
Petofi, G.	International Atomic Energy Agency
Stevenson, J.	Canadian Nuclear Safety Commission, Canada
Synak, D.	Slovenské Elektrárne, a.s., Slovakia

Technical Meeting

Vienna, Austria: 9–12 November 2021

Consultants Meetings

Vienna, Austria: 10–12 November 2020, 16–18 February 2021, 13–15 July 2021

 **IAEA**
International Atomic Energy Agency

ORDERING LOCALLY

IAEA priced publications may be purchased from the sources listed below or from major local booksellers.

Orders for unpriced publications should be made directly to the IAEA. The contact details are given at the end of this list.

NORTH AMERICA

Bernan / Rowman & Littlefield
15250 NBN Way, Blue Ridge Summit, PA 17214, USA
Telephone: +1 800 462 6420 • Fax: +1 800 338 4550
Email: orders@rowman.com • Web site: www.rowman.com/bernan

REST OF WORLD

Please contact your preferred local supplier, or our lead distributor:

Eurospan Group
Gray's Inn House
127 Clerkenwell Road
London EC1R 5DB
United Kingdom

Trade orders and enquiries:
Telephone: +44 (0)176 760 4972 • Fax: +44 (0)176 760 1640
Email: eurospan@turpin-distribution.com

Individual orders:
www.eurospanbookstore.com/iaea

For further information:
Telephone: +44 (0)207 240 0856 • Fax: +44 (0)207 379 0609
Email: info@eurospangroup.com • Web site: www.eurospangroup.com

Orders for both priced and unpriced publications may be addressed directly to:

Marketing and Sales Unit
International Atomic Energy Agency
Vienna International Centre, PO Box 100, 1400 Vienna, Austria
Telephone: +43 1 2600 22529 or 22530 • Fax: +43 1 26007 22529
Email: sales.publications@iaea.org • Web site: www.iaea.org/publications